W9-CGV-338

Social Institutions and Economic Development
A Tribute to Kurt Martin

Social Institutions and Economic Development

A Tribute to Kurt Martin

edited by

Valpy FitzGerald

University of Oxford, United Kingdom and Institute of Social Studies, The Hague, The Netherlands

KLUWER ACADEMIC PUBLISHERS

DORDRECHT / BOSTON / LONDON

A C.I.P. Catalogue record for this book is available from the Library of Congress.

ISBN 1-4020-0894-5

Published by Kluwer Academic Publishers,
P.O. Box 17, 3300 AA Dordrecht, The Netherlands.

Sold and distributed in North, Central and South America
by Kluwer Academic Publishers,
101 Philip Drive, Norwell, MA 02061, U.S.A.

In all other countries, sold and distributed
by Kluwer Academic Publishers,
P.O. Box 322, 3300 AH Dordrecht, The Netherlands.

Printed on acid-free paper

Printed in the Netherlands

For Kurt,
Scholar, Mentor and Friend

Contents

Foreword

Jan Pronk

The role of institutions in economic development has been debated at length. It is a major chapter in the history of economic thought. It was also a key issue in comparisons of the effectiveness of Eastern and Western economic systems. Understanding the variety of social and cultural institutions has always been crucial in analysing development processes in Africa, Asia, the Middle East and Latin America. Less attention has been given to institutions in studies of the economic performance of Western countries. This may be because economic policies in the West were mostly oriented to the short and medium terms rather than to the long-term perspective. In the short run institutions are given, in the long run they lend themselves for change.

From the outset, economic institutions (e.g. markets, enterprises) and their underlying values (e.g. efficiency, economic freedom) received much attention. Similar attention was given to political institutions (the state, government, the law) and values (democracy, accountability, human rights). Thought also turned to social institutions (entrepreneurship, the middle class, the family household, land-tenure systems) and social values (tradition, gender and age relations, justice). Studies soon followed of cultural institutions (religion, ethnicity) and values (material consumerism or the bond between man and nature). Without the insight gained by studying institutions, economics would have become a dull discipline.

There have been periods in which economists focused so strongly on policymaking that they tended to neglect the role of institutions. In the West, the post-War period of Keynesian macroeconomic policymaking is such a period. In the South, the adjustment years of the 1980s provides a similar example. Both proper management of the business cycle and effective handling of structural imbalances are necessary to restore a path of sustained economic growth. However, both require good insight into the policy environment and thus into institutions. Uniform macro-policies, irrespective of history, culture, ecology and power relations, will fail to bring about stability and growth.

However, it is not enough to simply emphasize the need to change institutions. In the 1990s the emphasis on the need for good governance in developing countries was too general. It was often perceived as foreign pressure, which met resistance and led to new bottlenecks. After 1989 calls for a quick dismantling of the institutions of the centrally planned economies led to new instabilities. Part of the power shifted to groups who were even less accountable than those they replaced. In the West, the lifting of restraints to market behaviour is freeing the way for corporate greed, proving detrimental to public trust in the system.

Understanding a process of change requires analysis in multiple dimensions. Rendering a policy advice requires specific knowledge of time, place and conditions. What is true or good in one country may turn out to be false or bad in another. What was valid a decade ago may be unreasonable today. Institutions differ and change. This explains why the study of economic development is so attractive and why giving policy advice is such risky business.

Will this be less so in the future? Globalization has lessened the relevance of national frontiers and boundaries between markets. The driving forces behind this process demand more uniformity throughout the world: uniform market standards, uniform good government criteria, uniform legislation, and uniform consumer demand. Globalization means that differences in time and place have lost some of their meaning: the scale of economic operations can be easily overseen wherever one is, and at each moment of time information is available anywhere.

However, even a uniform global process of change will have different impacts in different environments. The large institutional variety at present, both within and between countries, will result in different responses to the call for uniformity. It is not a matter of global technological change simply overtaking national institutional rigidities. Institutions themselves are alive. They are man-made, the result of people's responses to past change. They can be adjusted to new change. Institutions are not the boundary conditions of change, only limiting the capacity for sustainable development. They are the results of change in the past and they sustain other results of that past change. They are also capable of giving further direction to change, partly by absorbing new challenges, partly by resisting them or limiting their effects, partly by canalizing them within society. Institutions can also do this when the driving forces behind new challenges come from outside.

So, globalization does not make the study of institutions less relevant. On the contrary, if we define institutions as ways and means by which people have tried to respond to challenges and master change, studying institutions is all the more relevant the more change (development, growth, progress, conflict, etcetera) is expected.

This book brings together a series of papers on institutions in economic development in honour of Kurt Martin. Kurt Martin (1904–95) was a renowned development economist who remained an active scholar until his death. The study of institutions was a major chapter in his work, a centrepiece. He was a pupil of the great classical economists at the turn of the nineteens into the twentieth century, and he lived through that century not only as an observer and analyst but also as an active citizen and a listening teacher. He did not belong to a particular school but he had many students, who were attracted by his capability not only to teach but also to learn, gaining new insights both from other disciplines and from the reality of institutional change. That makes Kurt Martin worth reading at this new turn of the century.

The authors of the papers in this book have done just that. They all benefited from Kurt Martin's contributions to economic thought. The papers contain analyses of change in institutions in different areas of economic development in different parts of the world. To try to understand the workings of institutions – man-made structures that gain a life of their own and thereby a capacity to guide a society – is a rewarding exercise, both for students of development and for policymakers. That makes this tribute volume not only a response to earlier thinking, but also a contribution to new thinking.

Jan Pronk

Introduction: Institutions in Modern Development Economics

Valpy FitzGerald

This book is based on a special lecture series entitled *Social Institutions and Economic Development* held at the Institute of Social Studies between 1999 and 2001. Kurt Martin, whose family kindly sponsored the lectures in his memory, was a professor of development economics at ISS in The Hague from 1969 until his retirement in 1985 – although he remained an active influence at the institute until his death in 1995 through both his frequent visits and the activities of the many ISS faculty members who regarded (and still regard) themselves as his alumni.

Born in 1904, Kurt was undoubtedly one of the most significant development economists of the twentieth century, and I was fortunate enough to be the amanuensis for his last book.[1] From him I began to understand the history of modern economic thought as being about structural transformation, where the trunk passes so to speak from the classical economists of the eighteenth and nineteenth centuries, through Central Europe and Russia in the early twentieth century and on to 'developing countries' in the latter half of the twentieth century. At the opening of the twenty-first century, the study of 'globalization' is the clear successor to this political economy tradition. From this perspective, it could even be argued that the neoclassical model that dominates current economic theory will eventually be seen as a 'deviant branch' from this common trunk.

It is frequently asserted – erroneously – that development economists underestimate the importance of institutions. While this may have been true of certain neoclassical economists, it was not a characteristic of the founding fathers of development economics.[2] Kurt Martin made institutions a central feature of his own work. He stressed their importance throughout the history of economic theory itself and also in the philosophical foundations of the subject.[3] He regarded Schumpeter, who is now considered a key figure in evolutionary economics, to be a dominant influence on early development economists.[4] Nonetheless, Kurt felt that economic institutions were the result of historically embedded processes of social conflict rather than simply a solution to problems of transaction costs and collective action dilemmas.

One of his colleagues at Manchester before Kurt came to The Hague was Nobel Laureate Arthur Lewis. Lewis dedicates a substantial part of his extremely influential book, which set the agenda for much of early development practice, to economic institutions.[5] Three topics that Lewis covers in some depth are 'the right to reward', 'trade and specialization', and 'economic freedom'; all of them very much in the neoclassical tradition of his fellow laureate Douglass North.[6] However, when Lewis goes on to examine the key 'development' cases of religion, slavery, the family and the organization of agriculture and cottage industry, he approaches an evolutionary definition of the process of institutional change in developing countries not far removed from the political economy of Kurt Martin, whose seminal text on industrialization Lewis freely acknowledges as having influenced his own views.[7]

The current interest in economic institutions is fuelled in part by a reaction to the failed reductionism of monetary stabilization policies and structural adjustment programmes, but also – and more significantly – by the realization that the process of globalization has favoured the countries with institutions most capable of coping with exogenous economic change. A global market has emerged as the result of the progressive liberalization of domestic economies, the diversification of investment and production and technological change.[8] But the dismantling of old institutions of national economic management (such as planning ministries and state enterprises) is clearly not enough. New institutions are also needed to support not only the market but also social infrastructure and productive learning.

The creation of a global market is, moreover, proceeding at differing speeds. The integration of capital markets is perhaps moving the quickest, but without the strict regulation that characterizes domestic financial markets in order to prevent systemic collapse and protect consumers. Further, there is more or less complete free trade in goods (with the striking exception of farm products) but not in services, because this requires regulatory harmonization. There is also almost complete immobility of labour *de jure*, even while *de facto* there are large illegal movements of labour and refugees. Moreover, this global market is *without institutions.* What appear to be market institutions, such as the World Trade Organization or the International Monetary Fund, are in fact intergovernmental bodies with neither the power to intervene directly in markets nor authority over families and firms. The absence of international commercial law in this context is even more surprising, because throughout the history of capitalism markets have required a legal foundation.[9]

The Kurt Martin Lecture Series brought together a distinguished cast of international experts on social institutions and economic development; to each of these experts an ISS faculty member 'replied' with another paper.

These papers form the basis for this book.[10] In addition, I felt it would be appropriate to include an unpublished paper[11] on agrarian reform by Kurt Martin himself as a prelude, not only because he was the dedicatee of the lectures but also because his paper presents an elegant analysis of the interplay between markets and institutions, which concerns all the following papers. In Chapter 1, therefore, Martin addresses issues raised by the evidence of rural poverty on a large scale and of increasing landlessness in many parts of the developing world, even in countries where economic growth has been impressive by historical standards. He asks to what extent and under what conditions this situation can be redressed by agrarian reform.

Although since his writing (in the 1970s) agrarian reform has gone off the development agenda, so to speak, rural populations have continued to grow and, along with attempts to intensify cultivation, all governments are trying to generate off-farm jobs to absorb the natural increase in the agricultural labour force. However, attempts to combine policies of industrialization with the economic and social objectives of agrarian reform cannot go far or continue for long unless agriculture is producing growing marketable surpluses for sale to the non-farm population. Kurt took up this issue because it arose so often in the history of development and because it forms part of a wider set of problems, including the political demand for access to land, the effects of land reform on agricultural production and the distribution of income and the 'intersectoral' relationship between agriculture and industry. Martin concludes that the extent to which the economic and social objectives of agrarian reform can be attained within the framework of private ownership of land depends largely on the success of policies to hasten the end of the labour-surplus condition.

In Chapter 2, Richard Nelson of Columbia University picks up the tradition of evolutionary economic growth theory as a critique of neoclassical theory, which is 'blind' to technical progress and neglects the role of national *systems* of innovation as expressions of conscious human action. Nelson suggests that the Schumpeterian tradition of evolutionary economics is gradually converging with institutional economics as pioneered by John Commons and, more recently, theorised by North, the former by exploring technological change and the latter by addressing human interaction. Institutionalists could learn from evolutionary economists' understanding of path-dependency; while evolutionists could learn from the institutionalists' formal analysis. Nelson suggests that one way to approach such an integration of ideas is through the notions of 'productive routines' and 'social technologies' which characterize effective ways of solving problems so that institutions can be seen as supportive to rather than as 'constraints' on economic agents in the market. The nineteenth century development of both mass-production manufacturing in the United States and the synthetic dyestuffs industry in

Germany neatly illustrates Nelson's argument. In sum 'physical and social technologies co-evolve'.

Nelson's is an essentially benign view of economic progress – at least in the advanced countries. While recognizing the need for collective action, it does not examine the political interests that might promote or thwart such action. As Martin's own essay in this volume indicates, the choice of both physical and social technologies (i.e. agricultural mechanization and land tenure) reflects the relative power of rich and poor and may affect both the progress of industrialization and the incidence of poverty.

In Chapter 3, Hans Opschoor explores the ecological aspects of development from an explicitly evolutionary perspective and radically widens the agenda set out by Nelson. Opschoor argues that development occurs on the basis of activities involving both the natural and the social spheres, and it involves notions such as learning, anticipation, positive externalities and feedback. The present rector of the ISS suggests that a reformulation of development theory could fruitfully draw on both ecological and evolutionary economics. The issue of the link between economic growth and environmental degradation is a good example: sustained economic growth is not necessarily environmentally sustainable, nor will it automatically become sustainable. Sustainable development is most likely to be the reflection of deliberate environmental and developmental policies and policy-induced technological as well as institutional innovations.

In a wider sense, Opschoor argues that development takes place in a world in which adaptations and options are a matter of human design and societal implementation – in a more or less rapid, potentially goals-oriented and cumulative way. According to Opschoor, even values and preferences and changes in them over time can be analysed and understood through evolutionary methods. A perceptive analysis of the recent debate on the 'Environmental Kuznets Curve' underpins Opschoor's argument; although he does not remark on the fact that while the original Kuznets model was based on intuition rather than evidence,[12] it was at least soundly derived from the accepted concept of the changing balance of labour allocation between agriculture and industry. The 'environmental' version in contrast seems to rely on little more than the notion of pollution as an inferior good.

Opschoor concludes that addressing such issues requires an advance in analytical and anticipatory capacity so as to enable society to better understand and predict society-environment interactions and co-evolutionary change. He proposes a move towards integrated modelling and theory building that would include relevant parts of the ecological systems in which economic processes are embedded. In sum, development economics must open up to ecological as well as evolutionary approaches rather than rely upon the sub-discipline of 'economics' alone.

Although Kurt himself put little weight on environmental issues,[13] he would have been pleased with the way that Opschoor incorporates the views of the classical economists into the current concerns of environmental economics. However, the idea that the 'natural sphere' has its own dynamic would have appeared strange to him[14] – Darwinian at best and at worst a hangover from the Physiocrats. Nonetheless, economists whom Kurt admired, from Alfred Marshall to Ester Boserup, have argued persuasively that an ecological approach is a useful way to capture the dynamic relationship between conscious humanity and the natural world. Indeed, the contemporary concept of 'sustainable development' clearly relates the notion of 'development' to rich as well as to poor countries.

Kurt Martin always saw economics as a tool for bettering the human condition in general and for reducing poverty in particular. It is in this spirit that in Chapter 4, José Antonio Ocampo argues that recent contributions to economic thinking have provided useful means for understanding the frustrations that trends in policy-making have generated and, in turn, provided a basis for alternative policies to promote economic growth. The current Executive Secretary of the UN Economic Commission for Latin America and the Caribbean, Ocampo starts by looking at methodological issues and 'stylized facts' on growth and distribution in developing countries, drawing basically from the Latin American experience. He then turns to the dynamics of changing structures of production. Afterwards, Ocampo sets out a simple yet persuasive model of the linkages between such dynamics and overall economic and productivity growth from which he draws significant policy implications.

At an analytical level, Ocampo draws extensively from both the new and the old development literature to elucidate the central theme that growth is intrinsically linked to the dynamics of productive structures and to the particular institutions that are created to support it. The mix of dynamic productive structures and a supportive macroeconomic environment are, in this interpretation, the key to successful development. While the broader institutional context and the adequate provision of education and infrastructure are essential background conditions, they do not generally play a direct role in determining changes in the momentum of economic growth. The two crucial services that institutions provide are reducing information costs and solving the coordination failures that characterize interdependent investment decisions in a market economy.

His might thus seem to be an 'anti-institutionalist' view, except that his own outstanding work on the economic history of Colombia and on macroeconomic management in Latin America proves otherwise. Rather it is a question of whether it is the institutions as such or the policymakers within them (national, but also international), and the economic doctrines they espouse, that actually determine outcomes. The assumption that dynamic productive structures, and the particular institutions that support them, are the automatic result of market mechanisms in general and of the integration into the world economy in particular is demonstrably false – in the case of Latin America at least, according to Ocampo's persuasive argument. This view that directed state intervention is necessary in order to achieve economic development is one which Kurt Martin supported. Indeed,

> [t]oday, in spite of the sophisticated analytical techniques, and often highly esoteric specialisms characterising late twentieth-century economic science, the primary object of the exercise remains the same as it was in the seventeenth century: to provide national administrators and their responsible agents with the objective knowledge required to design and implement efficient economic policies (Deane 1989: v).

This theme of the conditions needed for sustainable economic growth in Latin America is continued in Chapter 5 by Rob Vos, currently deputy rector of the ISS. Vos points out that 'founding fathers' of development economics such as Paul Rosenstein-Rodan not only regarded exports as central to growth (they were thus not strict adherents to import substitution as their critics held) but they also saw export growth as more than a problem of overcoming low income elasticities of world demand. The more important point was the need to create technological externalities (such as 'learning by doing' and adequate social overhead capital) even in an open economy. In today's language, the core of endogenous growth capability is whether an economy possesses adequate infrastructure, human capital and entrepreneurial skill (including workers' learning by doing) to take advantage of opportunities provided by the global economy. Having put it this way, Vos is perhaps justified in asking, "What's new about the new growth theory?"

Vos nonetheless recognizes that the policy environment has changed dramatically since the post-War decades when development economics was first established. The founding fathers saw economic planning, aid and trade protection as important instruments to overcome the perceived development bottlenecks. Today's conventional wisdom is that globalization and free flows of commodities and capital are the prime movers of growth and development. However, summarizing his own careful quantitative research conducted in a large number of Latin American countries, Vos asserts that economic liberalization during the past three decades has not in fact brought the type of 'Big Push' that its advocates hoped for. Further, this 'post-modern' growth process seems to have exacerbated inequality rather than resolved it. In other words, institutions must be constructed not only to make production and investment more efficient, but also to reduce inequality and mass poverty if the resulting growth is to be sustainable. Needless to say, this view is entirely consistent with Kurt Martin's own view of the central dilemma of development.

Institutions are inhabited by people, not just individuals but also by groups of people bound together by various socially constructed ties. In Chapter 6, Frances Stewart of the University of Oxford provides an overview of the role of 'groups' with economic functions in development, drawing on both neoclassical and communitarian views. Her purpose is to explore different types of group behaviour and to identify reasons why some groups appear to work better than others. Given that humans invariably operate in groups, group behaviour is a systemic phenomenon which we must explore if we are to understand development. After defining some suitable categories for classifying group behaviour, Stewart presents case studies of public sector and community groups, contrasting successes and failures in several areas. The argument she advances does not intend to deny the important role of the market in resource allocation and innovation. Rather it emphasizes that collective action also has a critical role to play in development. If collective efforts are undermined by the norms promoted as a result of the enlarged role for the market and a diminished role for government, groups will still emerge but they will become more destructive than constructive of development.

Stewart regards groups as having both efficiency and claims functions: norms matter as well as the immediate interests of group members. Group behaviour is based on three models: power and control, incentives, and long-term reciprocity. The originality of Stewart's approach is the contention that group motivations are socially constructed and an essential part of the development process. They therefore cannot be reduced to contractual arrangements between members. The examples given, oriented logically enough to the resolution of poverty, are of cooperative ventures at the local and micro level and provide stimulating insights.

An institutional 'triad' of households, firms and governments forms the basis of modern economic theory. Each element has distinct objectives, constraints and powers: households save, consume, provide labour; firms invest, produce and employ labour; governments regulate the market, provide infrastructure and redistribute income. An essential part of the neoclassical project is to deny this institutional system: firms are reduced to extensions of shareholders' desires through the 'corporate veil'; families are seen as contractual arrangements between members; and governments are at best providers of services to voters and at worst a vehicle for achieving the personal aims of state elites. This seems to be essentially misguided, as Stewart's 'group principle' makes clear. Nonetheless, it is still far from clear why this triad of household, firm and government is so durable and why other types of group are so unstable.

In Chapter 7, FitzGerald[15] addresses the emerging model for development assistance in the new century on the assumption that the full integration of developing countries into the global economy requires correcting for failures in product and factor capital markets. 'Aid' as such would be confined to humanitarian emergencies and activities with large international externalities. However, without free labour movement, the logical implication of this approach is that citizenship itself becomes an economic asset. This model contrasts starkly with a second model: the notion of development assistance based on resource transfers to the most vulnerable countries or social groups as a form of 'social entitlement' in a global market. People's right to a minimal social entitlement can be derived from accepted principles of the theory of justice and are implied by international law. However, this model of global social citizenship also logically implies a corresponding system of international taxation. Despite the dominance of the first model due to the present geopolitical balance (or rather imbalance) of power, some trends in international political economy – such as repeated financial crises and outbreaks of armed conflict – may favour emergence of the second.

The difficulties of collective action, of how to form an effective regulatory grouping as a prelude to global institutions, are basically ones of political economy. Institutional economics does not provide us much help in resolving them. Even the meta-theory of foundational social contracts in the tradition of modern liberal political philosophy is not very useful. Indeed John Rawls' indifference principle is explicitly held to be inapplicable to global institutions – basically because the citizens of rich countries do not recognize the vast majority of the world population as party to a social contract that underpins the global economy.[16] While nation-states are founded on political constitutions, there is no international equivalent of their constitutions; even the United Nations is an *inter-state* body.

Inter-state relations between developed and developing countries are exemplified by the institutional structure of what is optimistically known as 'development cooperation' and more commonly as 'aid'. In Chapter 8, Ben Ndulu of Tanzania examines the reasons behind the poor record of aid effectiveness in Africa with particular focus on aid relationships as the key influence. His main hypothesis is that aid can be more effective if there is greater inclusiveness in the design and execution of aid programmes and strong citizen voice as a mechanism to enforce accountability and commitment to results. Ndulu reviews wisdom gleaned from his long and broad policy experience in Africa on what matters most for aid effectiveness. This leads him to propose a conceptual framework for an effective aid relationship based on effective accountability systems for achieving the results desired and for enforcing broad-based preferences.

However, if the proposed 'partnership approach' to aid relationships is to have any real meaning, other than as a public relations gloss on traditional inter-state dependence, then this framework must be implemented by donors in the conscious realization that they will have to cede power in exchange for effectiveness. Operationalizing a partnership approach will be far from easy, although Ndulu makes some valuable suggestions on how it might be achieved. In particular, he stresses the need for an explicit contractual relationship between donor and recipient, greater transparency and participation in the negotiation process and institutional reform in the private sector (both non-governmental organizations and business) as well as in the public sector. This is not only a matter of ensuring more efficiency and less waste, but also of generating a stronger political constituency for sustained development.

Ndulu's is still a relatively 'instrumental' view of the operations of aid institutions; it assumes a common view of objectives and thus a shared economic discourse. However, as Kurt Martin frequently pointed out, such assumptions should never be taken for granted. In Chapter 9, Marc Wuyts of the ISS explores what he terms the historic conceptual reversal in economic discourse 'from poverty to unemployment' then 'from unemployment back to poverty'. He links this latter transition with its accompanying changing emphases in foreign aid – that is, with the transition from aid as investment support to aid as poverty alleviation. Wuyts argues that the reversal from 'unemployment back to poverty' needs to be situated within the broader process of a 'spiral reversal' in capitalist development from the 1980s onwards. The approach adopted in this chapter, like that of Kurt Martin himself, is rooted in classical political economy and thus is a fitting closure to the volume. Wuyts draws inspiration not only from the literature of the early pioneers of development economics (including Kurt Martin himself) and the debates surrounding aid and structural adjustment policies, but also from the recently emerging literature on the history of the theory and practice of social statistics.

Wuyts reviews the earlier transition from poverty to unemployment as the central concern of economic discourse for public policy, in England in particular and in the industrialized world in general, during the late nineteenth and early twentieth centuries. He then turns to development economics proper and looks at the reverse transition from unemployment and absorption of the labour surplus to poverty as the key concern of economic policy in developing countries in the late twentieth century. Wuyts concludes by suggesting that the changing emphases in foreign aid – the transition from 'aid as investment support' to 'aid as poverty eradication' – reflects a broader process of transition from the centrality of the employment relation back to poverty.

Kurt Martin reflected in his own life the grand tradition of European social thought since the Enlightenment, the contradictory combination of economic rationality, humanism and the desire for justice.[17] He was a natural socialist, but already highly critical of 'actually existing socialism' in the early years of the last century. The demise of explicitly socialist states at the end of the century was nonetheless a disappointment because he held that modern civilisation should be based on the principles of cooperation rather than those of competition. Capitalism, however productive and even creative, is also unjust and wasteful, and thus ultimately irrational. In this sense Kurt was very much a modernist child of his time, a *pre*-1914 economist. He never became a post-modernist and was justifiably proud of having belonged to the early Frankfurt School of political economy rather than the later – and more famous – school of critical theory.

These lectures – and this book – thus continue the kind of debate that Kurt used to enjoy on the themes he thought were important. In the best sense, therefore, this is a fitting tribute to an extraordinary man, an influential scholar and a much-loved friend. Nonetheless, I am sure that had he been here for these lectures, he would have pointed out, as he so often did in the coffee room of *Die Wittebrug*, politely yet firmly that all of our eloquent theorising about institutions and economic development had already been said "so much more clearly by Ricardo and Marx".

Notes

1. That is, Martin (1991).
2. See the introductory chapter of Cooper and FitzGerald (1989).
3. He always recommended Schumpeter on the history of economic thought (Schumpeter 1954) to students, although he himself preferred Lowe on economic epistemology (Lowe 1965).
4. Particularly Schumpeter (1934) originally published in German in 1912, rather than the better-known Schumpeter (1942) to which evolutionary economists usually refer.
5. Lewis (1955) Part III.
6. For instance, North (1990) sees the 'demand for' and 'supply of' institutions interacting in a progressive manner over time.
7. Mandelbaum (1945).
8. Turning the end of the Cold War into a historical time warp, as if we were leaping backwards to 1914.

9. See Chapter 8 in this volume.
10. Refer here to the 'le Nez' lecture and the Saith reply; neither were written up and thus are not in this volume.
11. In fact this paper initiated the well-known *ISS Working Papers Series* in 1981.
12. As Martin (1991: 47) points out.
13. "Ecology as far as I can see has been neglected. An Indian author once said that poverty is our real pollution" (Martin, 1991: 52).
14. Unfortunately they did not have the opportunity to debate these issues.
15. As editor, I have to adopt the rather contrived position of referring to myself in the third person. Kurt Martin had the advantage of being able to refer to himself as 'Mandelbaum' when writing about his earlier work.
16. See Rawls (1999) for instance; although he overlooks the fact that the citizens of developing countries are offered little choice as to the rules upon which they wish the world economy to be organised. But then the slaves had little input to the US constitution either.
17. For a short intellectual biography, see my essay 'Kurt Mandelbaum and the Classical Tradition in Development Theory' which provides the opening chapter to Martin (1991).

References

Cooper, C. and E. V. K. FitzGerald (1989) *Development Studies Revisited: Twenty-Five Years of the Journal of Development Studies.* London: Cass.
Deane, P. (1989) *The State and the Economic System: An Introduction to the History of Political Economy.* Oxford: Oxford University Press.
Lewis, W. A. (1955) *The Theory of Economic Growth.* London: George Allen & Unwin.
Lowe, A. (1965) *On Economic Knowledge: Toward a Science of Political Economics.* New York: Harper & Row.
Mandelbaum, K. (1945) *The Industrialization of Backward Areas.* Oxford: Blackwell.
Martin, K. (ed.) (1991) *Strategies of Economic Development: Readings in the Political Economy of Industrialisation.* Basingstoke: Macmillan.
North, D. (1990) *Institutions, Institutional Change and Economic Performance.* Cambridge: Cambridge University Press.
Rawls, J. (1999) *The Law of Peoples.* Cambridge, MA: Harvard University Press.
Schumpeter, J. A. (1934) *The Theory of Economic Development.* Cambridge: Cambridge University Press.
——— (1942) *Capitalism, Socialism and Democracy.* New York: Harper & Row.
——— (1954) *History of Economic Analysis.* New York: Oxford University Press.

1 Agrarian Reforms and Intersectoral Relations: A Summary

Kurt Martin

Introduction

There is ample evidence of large-scale rural poverty and increasing landlessness in many parts of the Third World, even in countries where economic growth has been quite impressive by historical standards. The present paper asks *to what extent and under what conditions this situation can be redressed by agrarian reforms,* or more specifically by land reforms – that is, by attempts to transform the agrarian structure by altering the distribution of land and the terms on which land is held and worked.

History offers many examples of different types of land reforms, but it is not my intention to give a historical survey or empirical descriptions. Nor shall I enter into a debate on the merits or otherwise of collectivization, as practised in communist countries. Rather, I'll try to build up and present a theoretical argument based on an interpretation of the experience of some Third World countries.

Economic and social objectives of land reforms[1]

My first point must be to say that there is no universally valid answer to the question raised above. Much depends on 'initial conditions', which are very different in different parts of the world. I am thinking here in particular of the person-to-land ratios in agriculture, which as a result of history vary widely across different regions. In much of Asia, for instance in the countries of the Indian subcontinent and on Java, there is so little land relative to the number of people in agriculture that no land redistribution could provide holdings of a viable size to the entire agricultural population, including the landless and near-landless households. Indeed, in India the landless are estimated to account for about a third of all agricultural families, and on Java the proportion is probably only slightly less. Under such conditions an equal distribution of all agricultural land would leave everyone with plots too small to support a family. Thus, whatever the merits of land reforms may be (see later), they will not by themselves resolve the critical problems of landlessness.

V. FitzGerald (ed.), Social Institutions and Economic Development, 1–8.

The situation is quite different in land-rich Latin America, where it is said that about 60% of the agricultural population, which are attached to the traditional latifundia-minifundia complex of agriculture, suffer problems of land shortage due solely to the high concentration of ownership. Given the abundance of land, the worst rural poverty here could be overcome by a redistribution, in which part of the properties of the very large landowners would be transferred to the poorest segments of the agricultural population, moving everyone (including the landless) nearer to the average ratio of labour to land. To my knowledge, no such reform has ever been implemented in Latin America, nor am I arguing that this would be the most sensible arrangement.

Although in the Asian countries where land has become scarce it *is* impossible to convert landless labourers into owner-cultivators, small farmers now living near or below the poverty line could obtain some additional land and become more viable, if effective (and possibly draconian) ceilings could be imposed on the size of holdings. It is generally held that this would promote productive efficiency as well as equity. An inverse relation has been observed – particularly in rice cultivation, less certainly in wheat – between the size of plots and yield per acre. This inverse relation is explained by the fact that small cultivators make more intensive use of land than large land owners do.[2] In Sri Lanka an average of two acres of irrigated rice land is considered sufficient to support a family of five or six, and for India the average figure quoted is only marginally higher. Neither are small cultivators held to be at a disadvantage when it comes to introducing technical innovations. The 'Green Revolution' technologies, for instance, are said to be 'scale neutral' in the sense that the response of output to the divisible inputs of water, fertilizers and seeds does not depend on the size of plots.

The above statements, which assert the superiority of an agriculture consisting of smallholders where the dominant unit of production is the family farm, need to be qualified for two reasons. First, there are certain activities in agriculture which always involve a scale of operation much larger than the individual holding. Many land improvement projects such as water management, flood control or land levelling can be carried out effectively only by planning at a village level. Yet when land is privately owned and operated, communal activities are difficult to organize, because some private owners will always gain more from such activities than others. Difficulties of this kind tend to impede cooperative and group farming in the mixed economies of the Third World, though there are variations from case to case because of differences in traditions, and much depends on the role of the state.

Second, is the 'scale neutrality' of land-saving modern technologies. Even if scale neutrality as applied to rice cultivation makes some technical sense, it is obvious that the inputs of the Green Revolution call for financial resources beyond the reach of the poorer peasants. Now considering that in a private enterprise economy no redistribution of land can be fully equalizing – that is, it can never eliminate all disparities in the size of holdings – land reformers invariably insist that credit facilities and other services be supplied in an equitable way to the (post-reform) small units so as to put them on par with the wealthier and stronger peasants. But that is demanding a lot. For reasons too well known to require elaboration, the supply of services to agriculturalists by governments or other organizations has tended to *increase* inequalities, because access to such non-land resources is almost always biased in favour of the more prosperous farmers. This trend cannot easily be reversed – it really is 'in the nature of things' – unless the rural poor are themselves mobilized as a force of change.

Where land is rented out to tenant-farmers or sharecroppers, agrarian reforms generally include measures for improving the terms and conditions of the tenure contracts. More radical are the reforms that confer property rights on all tenants and thus abolish the social organization of agriculture based on absentee (rentier) landlordism. Such reforms were carried out in the Indian state of Kerala, for instance, in Taiwan and, with some exemptions, in South Korea (where, however, tenancy seems to have reappeared on a certain scale). The effect of such reforms should be to stimulate food production, which tends to be rather static in a system of agriculture dominated by a class of absentee landlords. Inequality is also likely to be reduced, although the category of tenants who benefit often includes a group of privileged agriculturists who controls part of the land under lease arrangements (owner-cum-tenants). It is also clear that such a reform does not directly affect the position of the class of landless labourers; it is a middle group of 'progressive' farmers who are likely to gain most.

In sum, while agrarian reforms of one type or another, if effectively implemented, have merit, they cannot by themselves solve the central problem of many Asian countries, where too many people are trying to live on too little land and where populations are growing fast. These conditions are bound to set up strong polarizing forces that continue to operate even in a reformed agriculture where small farmers have obtained some additional land. For although these reforms can have the great advantage of increasing the intensity of cultivation and the degree of utilization of family labour, their effect on the demand for hired labour is at best uncertain,[3] while landlessness is likely to grow under continuing demographic pressure. A satisfactory solution to the problems that lead to demands for agrarian reforms thus depends on the development strategy adopted for the economy as a whole; it cannot be found in attempts to tackle the problems within the agricultural sector alone.

The labour-surplus condition

More concretely, in countries where agriculture is overcrowded, while the population continues to grow the achievement of lasting improvements in the social conditions of the agricultural labour force depends very much on the success of attempts to hasten the end of the labour-surplus condition. Along with greater intensity of cultivation such attempts must include programmes of sustained industrial growth so that regular off-farm work is provided for a gradually rising proportion of the natural increase in agricultural labour force.

However, attempts to combine policies of industrialization with the economic and social objectives of agrarian reforms can create a number of special problems, and perhaps dilemmas, to which we must draw attention. These problems involve the interaction between the different sectors of the economy.

Industrial growth cannot go far or continue for long unless agriculture produces growing marketable surpluses for sale to the non-farm population, which in this way obtains the food and agricultural materials needed for industrial expansion. Now, it has been argued that agrarian reforms in favour of smallholders leave little scope for growth of the marketed surplus of agriculture, because smallholders generally devote a higher proportion of output to their own consumption than the larger units do. Consequently, the growth of industry might slow or come to a halt due to the lack of essential supplies. This problem has a long history – it was at the centre of the Russian industrialization debate in the 1920s.

But this is not the place for a lengthy discussion or historical reflection. Let me merely make two brief comments. First, some of the agrarian reforms in the market economies of the Third World – for example, the reforms in India – led to the elimination of intermediaries rather than to far-reaching redistribution and equalization of holdings among cultivators. Such reforms have not been an obstacle to the continued existence or emergence of a group of surplus-producing big farmers (*kulaks*). So although it is true that lack of progress in agriculture has often limited industrial growth, one can hardly say in these cases that the limits were set by the agrarian reforms. Second, several studies show that the land reforms carried out after World War II in countries such as Egypt, Taiwan and South Korea were accompanied by a strengthening of state control over various aspects of agriculture. The purpose of these controls and of the activities of state-controlled institutions (supervised cooperatives, procurement agencies) has been to try to ensure that the reformed agriculture consisting mainly of smallholders functions in a way that fits the overall government strategy. In fact of Taiwan it was said that after land reform the state took the place of the landlords in extracting surpluses from agriculture.[4]

In this context the surplus said to be 'extracted' from agriculture is different from the marketable surplus both conceptually and in magnitude. The example of Taiwan above refers to a financial or 'investable' surplus of agriculture, which is available for transfer to other sectors of the economy where it can be used to finance the growth of non-agricultural activities. For such a net transfer of financial resources out of agriculture to occur, the proceeds from the sale of agricultural produce to the non-farm population (and to foreign countries) must be in excess of the 'import' of goods and services *into* agriculture.

There are three transfer mechanisms. The transferable surplus may arise on private accounts, if the monetary savings of agriculturalists are larger than private investment in agriculture (some of which may come from non-agricultural sources). It may arise on public account, if governments *via* direct and indirect taxation withdraw more income from agriculture than they spend on agricultural projects and subsidies. Moreover, quite apart from the budget, governments through their trade policies and price regulations for outputs and inputs can affect the terms of trade of agriculture. If these terms deteriorate, there is an invisible resource outflow from agriculture, even if the quantities traded remain unchanged. Governments have often used this power to bring about an outflow of funds from agriculture that would not have occurred otherwise.

It can be argued that at low levels of industrial development such capital transfers from agriculture may have a certain rationality, particularly in the absence of foreign capital inflows or when no funds for industrial development are available from profits from other primary activities such as mining or oil (see later). For when the starting base of industry and other surplus-producing non-agricultural activities is still very small, an industrial expansion based solely on the internal savings of the 'modern' sectors is bound to be very weak in *relation to the supply of labour* to industry. Indeed, this supply is practically 'unlimited' where there is high and growing pressure of labour on land. A capital contribution from agriculture can make a difference then, although not necessarily a major or decisive one. Even then there are times when a combination of circumstances leads to a temporary inflow of funds *into* agriculture, for instance, when major irrigation works are undertaken. But in general it seems that during early industrialization the flow of financial resources is in the reverse direction.[5]

It is necessary to add that enforced resource transfers from agriculture have often been economically counter-productive and socially regressive or even parasitical. Indeed it seems obvious that to make the peasants foot the bill for industrialization must conflict with the objectives of agrarian reforms and with the aim to alleviate rural poverty. Thus the situation is full of contradictions and dilemmas, which in one way or the other have to be resolved.

Historical experience shows that these conflicts are reduced if the resource outflows from agriculture occur along with rising agricultural productivity. The two processes can go together, provided that the productivity gains in agriculture do not themselves necessitate large-scale capital investment *within* agriculture – and we know that they often do not require that.[6] The economic and social implications of financial outflows from an expanding agriculture, where productivity and incomes are rising, are clearly very different from a squeeze on a stagnating agriculture.

Much also depends on the pattern of industrial growth and hence on the use that is made of the investable funds (some of which may come from agriculture). Suppose that a relatively high proportion of these funds goes into the establishment of agricultural (or rural) industries. These are industries which through backward or forward linkages are geared to the needs of a developing agriculture so that their establishment stimulates agricultural progress. Moreover, many of these industries can be set up in rural areas or small towns. Such rural industrialization involves an intersectoral transfer of labour – a shift of labour from agriculture to industry (or from agricultural side-occupations to industry proper) – *without migration to the cities*. Since the labour force of these industries largely consists of people who remain members of agricultural households, the income of farm families is increased and the gap between urban and rural earnings is narrowed. To achieve this usually calls for a degree of state involvement in the economy, and one of the purposes or effects of such involvement may be to secure a capital contribution from agriculture. However, regardless of whether there is such surplus 'extraction' from agriculture, some of the conflicts of interest between and within urban and rural groups will lose force if industrialization takes the course described (and if in a reformed agriculture a more egalitarian pattern of landholding has been established).

Conclusion

Outflows of funds from agriculture are not by any means a universal requirement for 'development'. For example, we all know that Indonesia is in the fortunate position of having large oil revenues (and much foreign aid) which can contribute to finance industry or agriculture; moreover, there are the plantations which would demand separate treatment. Indonesia therefore has no need to try to obtain funds for industrial growth from the peasantry. All the same, I have taken up this particular issue because it has often arisen in the history of development, and where it did arise, it formed part of a wider set of problems of a more general nature. In listing some of these problems I started with the demand for agrarian reforms and the possible effects of such reforms on agricultural production and the distribution of income. But as agriculture does not exist in isolation, I had to give some attention to intersectoral relations.

I confined myself to the situation that exists where there is heavy pressure of labour on land, ignoring the possibilities of collectivization, as practised in most communist countries. On this basis my main argument was that the extent to which the economic and social objectives of agrarian reforms can be attained (within the framework of private ownership of land) depends largely on the success of policies which can hasten the end of the labour-surplus condition. These policies call for a variety of measures across the economy. Many of these measures are country-specific, but they must always include both an increase in the intensity of cultivation and a sustained expansion of regular non-agricultural employment opportunities, particularly in rural areas. Dilemmas of policy and conflicts over priorities are bound to arise, and the way they are resolved will depend on the nature of the political system and of the coalition of social forces which dominates decision-making.

Notes

1. This section contains, in passing, some brief references to land reforms in India as a whole, in the Indian state of Kerala, in South Korea and in Taiwan. The following studies are particularly useful for a general overview: Rudra (1978), Herring (1980), Ranis (1978) and Lee (1979).
2. Two other factors which have nothing to do with 'efficiency' operate in favour of the small family farm. First, the small unit enjoys the cost advantage of being able to employ unpaid (or cheap) family labour, although in rice cultivation even small farmers often employ some hired labour for transplanting and harvesting. Second, the family labour force has the unique ability to combine agricultural work with all manner of side occupations.
3. On this point see Bell and Duloy (1974).
4. For this see Abdel-Fadil (1975) and Radwan and Lee (1979).
5. On this topic there are a number of interesting studies which are comparable, although they differ slightly in definitions and coverage. I would like to mention Lee (1971), which is discussed with references to other cases in Mellor (1973); Mundle and Ohkawa (1979), Sharpley (1979) and Lipton (1978). Also of interest is the study by M. Ellman (1975) on whether agricultural surplus provided the resources for increased investment in the Soviet Union during its first five year plan.
6. In technical jargon, our text implies that financial resource outflows from agriculture and agricultural productivity gains can go together provided that the marginal capital-to-output ratio in agriculture is less than one. This is often

found to be so in the less-developed countries. What is relatively costly (apart from major irrigation works such as the Aswan Dam) is the provision of rural infrastructure, such as rural electricity, transportation and marketing facilities. But *these are non-agricultural activities and investments*, which serve agriculture from outside it: they bring 'external economies' to the farmers (as well as to non-farmers in rural areas).

References

Abdel-Fadil, M. (1975) *Development, Income Distribution and Social Change in Rural Egypt 1952–70*. Cambridge: Cambridge University Press.

Radwan, S. and E. Lee (1979) 'The state and agricultural change: A case study of Egypt 1952–1977', in: Dharam Gai et al., *Agrarian Systems and Rural Development*. London: Macmillan.

Bell, C. L. G. and J. H. Duloy (1974) 'Rural target groups', in: H. B. Chenery et al., *Redistribution with Growth*, pp. 113–135. Oxford: Oxford University Press.

Rudra, A. (1978) 'Organisation of agriculture for rural development: The Indian case', *Cambridge Journal of Economics*, 2 (4): 381–406.

Herring, R. J. (1980) 'Abolition of landlordism in Kerala', *Economic and Political Weekly (Bombay), Review of Agriculture*, 15 (26) June: A59–A69.

Ranis, G. (1978) 'Equity with growth in Taiwan: How special is the special case?', *World Development*, 6 (3): 397–410.

Lee, E. (1979) 'Egalitarian peasant farming and rural development: The case of South Korea', *World Development*, 7 (4/5): 493–517.

Lee, T. H. (1971) *Intersectoral Capital Flows in the Economic Development of Taiwan 1895–1960*. Ithaca, NY: Cornell University Press.

Mellor, J. W. (1973) 'Accelerated growth in agricultural production and the intersectoral transfer of resources', *Economic Development and Cultural Change*, 22 (1): 1–16.

Mundle, S. and K. Ohkawa, (1979) 'Agricultural surplus flow in Japan, 1888–1937', *The Developing Economies*, 17 (3) September: 247–265.

Sharpley, J. (1979) 'Intersectoral capital flows, evidence from Kenya', *Journal of Development Economics*, 6 (4): 557–571.

Lipton, M. (1978) 'Transfer of resources from agriculture to non-agricultural activities: The case of India', in: J. F. J. Toye (ed.), *Taxation and Economic Development*. London: Frank Cass.

Ellman, M. (1975) 'Did the agricultural surplus provide the resources for the increase in investment in the U.S.S.R. during the first five year plan?' *Economic Journal*, 85 (340) December: 844–863.

2 Bringing Institutions into Evolutionary Growth Theory[1]

Richard R. Nelson

The economists who have been active in the development of evolutionary growth theory over the last twenty years were motivated in large part by their perception that neoclassical economic growth theory, while assigning technological change a central role in economic growth, is totally inadequate as an abstract characterization of economic growth fuelled by technological change (Nelson and Winter 1982). In particular, neoclassical theory represses the fact that efforts to advance technology are to a considerable extent 'blind'. This proposition does not deny the purpose, the intelligence and the often powerful body of understanding and technique that those aiming to advance technology bring to their work. But it always seems to be the case that different inventors and R&D teams lay their bets in different ways, and what will work best is virtually impossible to predict in advance.

Hence industries and eras with rapid and cumulative technological advance have almost always been marked by a number of competing efforts and actors, with ex post selection rather than forward-looking planning determining the winners and losers. The broad notion that technological advance proceeds through an evolutionary process has been developed independently by scholars operating in a variety of different disciplines. These include sociologists (Constant 1980, Bijker 1995), technological historians (Rosenberg 1976, Vincenti 1990, Petroski 1992, Mokyr 1990) and economists interested in modelling (Nelson and Winter 1982, Metcalfe 1998, Saviotti 1996).

Needless to say, explicit recognition that technological advance progresses through an evolutionary process leads one to formulate a growth theory with a very different structure than that of neoclassical growth theory, new or old. However, for the most part evolutionary growth theory, like neoclassical growth theory, has yet to take on board the complex institutional structures that characterize modern economies.[2]

On the other hand, sophisticated empirical scholars of technological advance have always understood that the rate and character of such advance is influenced by the institutional structures supporting it. Moreover, institutions strongly condition whether and how effectively new technology is accepted and absorbed into an economic system. These themes are clear, for

V. FitzGerald (ed.), Social Institutions and Economic Development, 9–21.

example, in David Landes' magisterial *Unbound Prometheus* (1970) and in Christopher Freeman's *The Economics of Industrial Innovation* (1982). More recently the notion of a national or a sectoral innovation system, which clearly is an institutional concept, has played a significant role in theorizing about technological advance (see e.g. Lundvall 1992, Nelson 1993, Mowery and Nelson 1999).

However, it seems fair to say that by and large modern evolutionary economists writing about technological change and modern economists stressing the role of institutions in economic development have had little interchange. The principal purpose of this and kindred essays is to build a bridge between the two intellectual traditions, and to suggest a way they may be joined.

Institutional analysis and evolutionary economic theory: The historical connections

I want to begin by proposing that, before modern neoclassical theory gained its present preponderant position in economics, much of economic analysis was both evolutionary and institutional. Thus, Adam Smith's analysis (Smith 1937 [1776]) concerned how "the division of labor is limited by the extent of the market". In particular, Smith's famous pin-making example certainly fits the mould of what I would call evolutionary theorizing about economic change. Indeed, his analysis is very much one about the co-evolution of physical technologies and the organization of work, with the latter, I would argue, very much a notion about 'institutions'. In many places in *The Wealth of Nations*, Smith is expressly concerned with the broader institutional structure of nations in a way that is certainly consonant with the perspectives of modern institutional economics. Karl Marx of course was both an evolutionary theorist and an institutional theorist. So too was Alfred Marshall if we consider the broad span of his writings. Thus, evolutionary growth theorizing that encompasses institutions in an essential way has a long and honourable tradition in economics.

As neoclassical theory gained dominance in economics – and increasingly narrowed its intellectual scope – both the institutional and the evolutionary strands of economic analysis became counter-cultures. In some cases they were intertwined, as in Veblen (1915, 1919) and Hayek (1967, 1973). The tendency, however, was for the dissonant strains of institutional economic theorizing and evolutionary economic theorizing to take their own separate paths. Thus, in the United States, Commons (1924, 1934) helped to define the American institutional school. His analysis was not very evolutionary however. Nor was the perspective of Coase (1937, 1960), who later played a major role in shaping 'the new institutional economics'. On the other hand, Schumpeter (1942), whose work arguably provided the starting point for modern evolutionary economics, is seldom footnoted by self-professed institutionalists, despite the fact that he was much concerned with economic institutions. And Schumpeter's institutional orientation was also ignored in early writings by the evolutionary economists who had cited him as their inspiration.

Thus, what has been called the 'new institutional economics' and the new evolutionary economics have different immediate sources, and their focal orientations have also differed. Institutional economics is oriented towards the set of factors that mould and define human interaction, both within organizations and between them. In contrast, much of modern evolutionary economic theorizing is focused on the processes of technological advance.

However, in my view at least, recent developments have seen the two strands coming together again, as Hodgson (1988, 1993) and Langlois (1989) have long argued should be the case. Thus, Douglass North (1990), perhaps today's best known economic 'institutionalist', gradually has adopted an evolutionary perspective regarding how institutions form and change. And, as I noted earlier, many of the scholars who did the early work on the new evolutionary economics recently became focused on such subjects as 'national innovation systems', which is an institutional concept par excellence.

There certainly are strong natural affinities in the form of common core assumptions and perceptions between institutional economists, at least those in the school of North, and modern evolutionary economists. There also are strong reasons more generally on why they should join forces.

Both camps share a central behavioural premise that human action and interaction should be understood as largely the result of shared habits of action and thought. In both there is a deep-seated rejection of 'maximization' as a process characterization of what humans do. They also reject the Friedmanian notion that, while humans do not go through actual maximizing calculations, they behave 'as if' they did, and therefore, that behaviour can be predicted by an analyst who calculates the best possible behaviour for humans operating in a particular context. Thus, for scholars in both camps, patterns of action are understood in behavioural terms, with improvements over time explained by processes of individual and collective learning. For economic evolutionary theorists, this exactly defines the nature of an evolutionary process.

Scholars in both camps increasingly share a central interest in understanding the determinants of economic performance and how economic performance differs across nations and over time. Modern evolutionary theorists focus centrally on what they tend to call 'technologies'. For them, a country's level of technological competence is seen as the basic factor constraining its productivity, with technological advance the central driving force behind economic growth. As noted, increasingly evolutionary economists are coming to see 'institutions' as moulding the technologies used by a society and technological change itself. However, institutions have yet to be incorporated into their formal analysis.

On the other hand, institutional economists tend to focus specifically on these institutions. Many would be happy to admit that a central way that institutions affect economic performance is through the influence of a country's institutions on its ability to master and advance technology. However, institutionalists have yet to include technology and technological change explicitly into their formulation.

The arguments for a marriage I think are strong. Below I map out what such a marriage might look like.

Routines as a unifying concept

I begin by noting the essential function that the notion of a 'routine', or an equivalent concept, plays in modern economic evolutionary theory. As Sidney Winter and I developed the concept (1982), carrying out a routine is 'programmatic' by nature. Like a computer program a routine tends largely to be carried out automatically. Like such a program, our routine concept admits choice within a limited range of alternatives, that is, a channelled choice.

Thus the routines built into a business firm, or another kind of organization that undertakes economic activity, largely determine what it does under the particular circumstances it faces. The performance of that firm or organization are determined by the routines it possesses and the routines possessed by other firms and economic units with which the firm interacts, including competitors, suppliers and customers. At any given time, many of the routines are common to most firms in the same line of business, but some are not. These latter provide the stuff that determines how firms do relative to their competitors. The distribution of routines in an economy at any given time determines overall economic performance. Under evolutionary economic theory, economic growth results from changes in the distribution of operative routines, associated both with the creation of superior new routines and with the increasingly widespread use of superior routines and the abandonment of inferior ones. The latter can occur through the relative expansion of organizations that do well or by the adoption of better techniques by organizations that had been using inferior ones, or both.

As noted, most writings by evolutionary economists have focused on 'physical' technologies as routines. However, the notion of a routine fits well with the conceptualization of many institutional economists, if the concept is turned to characterize standardized patterns of human transaction and interaction more generally. Indeed, if one defines institutions as widely employed 'social' technologies, in the sense I will develop shortly, it is easy to take institutions on board as a component of an evolutionary theory of economic growth.

In order to see what I am suggesting here, it is useful to reflect a bit on some important characteristics of productive routines. A routine involves a collection of procedures which, together, result in a predictable and specifiable outcome. Complex routines almost always can be analytically broken down into a collection of subroutines. Thus, the routine for making a cake involves subroutines like pour, mix and bake. These operations often require particular inputs, like flour and sugar and a stove. In turn, virtually all complex routines are linked with other routines that must be effected in order to make them possible, or to enable them to create value. Thus, a cake-making routine presupposes that the necessary ingredients and equipment are at hand, and the acquisition of these at some prior date requires its own 'shopping' routines. Still further back in the chain of activity, the inputs themselves needed to be produced, in a form that meets the requirements of cake makers.

A key aspect of productive routines is that, while the operation of a particular routine by a competent individual or organization generally involves certain idiosyncratic elements, at its core there are almost always elements that are broadly similar to what other competent parties would do in the same context. By and large, the ingredients and the equipment used by reasonably skilled bakers are similar. And the broad outline of steps taken to make a cake can generally be recognized by someone skilled in the art as being roughly those described in *The Joy of Cooking* or some comparable reference.

There are two basic reasons why productive routines tend to be widely used by those who are skilled in the art. The first is that great cake recipes, or effective ways of organizing bakeries or producing steel or semiconductors, tend to be the result of the cumulative contributions of many parties, often operating over many generations. This is a central reason why they are as effective as they are. Widely used routines are widely used because they are effective; and they are effective because over the years they have been widely used. To deviate from them in significant ways is risky, and while the payoffs may be considerable, there also is a large chance of failure.

The second reason why particular routines are widely used is that these routines tend to form part of systems of routines. This systemic aspect forces a certain basic commonality of ways of doing particular things. The needed inputs tend to be available, routinely, for widely known and used routines. If help is needed, it generally is easy to find someone who already knows a lot about what is needed and to explain the particulars in common language. In contrast, idiosyncratic routines tend to lack good fit with complementary routines and may require their users to build their own support systems.

Social technologies and institutions

In an earlier paper (Nelson and Sampat 2000) in which Bhaven Sampat and I developed many of these notions, we proposed that, if one reflects on the matter, the program built into a routine generally involves two aspects: firstly a recipe that is anonymous regarding any division of labour and secondly a division of labour plus a mode of coordination. We proposed that the former is what scholars often have in mind when they think of 'physical technologies'. The latter we called a 'social technology' and proposed that social technologies are what many scholars have in mind when they use the term 'institutions'. North and Wallis (1994) proposed a similar distinction between physical and social technologies.

Widely employed social technologies certainly are defined by and define 'the rules of the game', which is the concept of institutions employed by many scholars. Social technologies can also be viewed as widely employed 'modes of governance', which is Williamson's notion (Williamson 1985) of what institutions are about. In the language of transaction costs that is employed in the institutional literature, widely used 'social technologies' provide low transaction cost ways of getting something done. As this discussion indicates, the concept of social technology is broad enough to encompass both ways of organizing activity within particular organizations – that is, the M form of organization is a social technology – and ways of transacting across organizational borders. Thus, markets define and are defined by 'social technologies'. So too are widely used procedures for collective choice and action.

This formulation naturally induces one to see prevailing institutions not so much as 'constraints' on behaviour, as do some analysts, but rather as defining effective ways to get things done when human cooperation is needed. To view institutions as constraints on behaviour is analogous to seeing prevailing physical technologies as constraints. A productive social technology (an institution) or a physical technology is like a paved road across a swamp. To say that the location of the prevailing road is a constraint on getting across is basically to miss the point. Without a road, getting across would be impossible, or at least much more difficult.

Institutions in an evolutionary theory of economic growth

The question of how institutions fit into a theory of economic growth of course depends not only on what one means by institutions, but also on the other aspects of that theory. I suggest that the concept of institutions as social technologies fits very nicely into evolutionary theories of economic growth.

Technological advance as the driving force

While these days almost all scholars studying economic growth see technological advance as a large part of the story, evolutionary theorists give special weight to technology. The reason is that, while neoclassical theory sees economic actors as facing a spacious choice set, including possible actions that they never took before, within which they can choose with confidence and competence, evolutionary theory sees economic actors as at any time bound by the limited range of routines that they have mastered. Each has only a small range of choice. Further, actors' learning of new routines is time consuming, costly, and risky. Thus while neoclassical growth theory sees considerable economic growth as possible simply by 'moving along the production function', in evolutionary theory there are no easy ways to master new things.

Put more positively, from the perspective of evolutionary theory, past economic growth is understood to be the result of the progressive introduction of new technologies, which were associated with increasingly higher levels of worker productivity and the ability to produce new or improved goods and services. As a broad trend, these technologies were also progressively capital using. Elsewhere (Nelson 1998) I have developed the varied reasons for the capital-using nature of technological change. Rising human capital intensity has been a handmaiden to that process, being associated both with the changing inputs that have generated technological advance and with the changing skill requirements of new technologies.

Within this formulation, new 'institutions' and social technologies come into the picture as changes in modes of interaction – new modes of organizing work, new kinds of markets, new laws, new forms of collective action – that are called for as the new technologies are brought into economic use. In turn, the institutional structure at any time has a profound effect on and reflects the technologies that are in use and those being developed.

I believe that the concept of institutions as social technologies, the routines language for describing them and the theory sketched above of how institutions and institutional change are bound with the advance of physical technologies in the process of economic growth, becomes more powerful the closer the analysis gets to describing actual social technologies in action. Thus I turn now to two important particular developments in the history of modern economic growth: the rise of mass-production industry in the United States in the late nineteenth century and the rise of the first science-based industry – synthetic dyestuffs – in Germany at about the same time. Given space constraints, the discussion must be sketchy, but I hope to provide enough detail to show the proposed conceptualization in action.

The rise of mass production

As Alfred Chandler (1962) and other business historians tell the story, in the late nineteenth century and the first half of the twentieth century, manufacturing industry in the United States experienced rapid productivity growth associated with new methods of production – that is, technologies or routines – that came to be called 'mass production'. Accompanying the rise of these methods were an increasing scale of plants and firms, rising capital intensity of production and the development of professional management, often with education beyond the secondary level. However, these increases in 'physical and human capital per worker' and in the scale of output should not be considered as independent sources of growth, in the sense of growth accounting; they were productive only because they were needed by the technologies being introduced.

At the same time, it would be a conceptual mistake to try to calculate how much productivity increase the new technologies would have allowed had physical and human capital per worker and the scale of output remained constant. The new production routines involved new physical technologies that incorporated higher levels of physical and human capital per worker than the older routines they replaced. To operate the new routines efficiently required a much larger scale of output than previously.

The new routines also involved new 'social technologies'. Chandler's great studies are largely about the new modes of organizing business that were required to take advantage of the opportunities for 'scale and scope'. The scale of the new firms exceeded that which owner-managers plus their relatives and close friends could deal with, either in terms of governance or finance. The growing importance of hired professional management and the diminished willingness of the original family owners to provide all of the financial capital required called for the development of new financial institutions and associated markets. The need for professional managers also brought business schools into being. More generally, the new industrial organization profoundly reshaped shared beliefs on how the economy worked and came to define modern capitalism.

The development of mass production proceeded especially rapidly in the United States, at least in part because of the large size of the American market, but also because the associated new institutions grew up rapidly in the New World. In general Europe lagged behind. On the other hand, the rise of new institutions to support science-based industry occurred first in Europe.

Synthetic dyestuffs

I turn now to consider the example of the rise of the first science-based industry, in Germany, which occurred at roughly the same time as the rise of mass production in the United States. Several scholars have told the basic story, but the account I draw on most here is that in the thesis by Peter Murmann (1998). Murmann presents his account in standard language. The account I present here is 'semi-formal' in the sense that it makes explicit use of the concept of routines and the physical and social technologies involved in routines.

Several new routines play a key role in the story. The first is a new 'physical technology' for creating dyestuffs with university-trained chemists as the key input. This new physical technology came into existence in the late 1860s and early 1870s as a result of improved scientific understanding of the structure of organic compounds. The second key element in the story is the development of the 'social technology' for organizing chemists to work in a coordinated way for their employer – that is, the invention of the modern industrial research laboratory. The third element in our story is another social technology: the system of training young chemists in the understandings and research methods of organic chemistry. This social technology was university based and funded by national governments. Finally, there are new markets with their own particular rules and norms. One links firms interested in hiring chemists with the supply of chemists. Another links dyestuff firms with the users of the new dyestuffs.

Several different kinds of 'institutionalized' organizations play a role in our theoretical story. First are the chemical products firms, which are of two types. The 'old' type possesses no industrial research laboratory and achieves new dyestuffs slowly through processes involving only low levels of investment. The other kind of firms, the 'new' type, invests in industrial research laboratories and, because of those investments, achieves new dyestuffs at a much faster rate than do old firms. There are two other kinds of organization in this story as well. One is national chemical products industry associations, which lobby government to support university training. The other is national universities, which train young chemists. National political processes and government funding agencies also are part of the story, but they are treated implicitly rather than explicitly.

As noted, this account also involves specification of certain 'institutionalized' markets and the recognition that these markets differ somewhat from nation to nation. In particular, chemists have a national identity, as do the firms. German chemists (assuming that these all are trained in German universities) require a significantly higher salary to work in a British firm than in a German one, and British-trained scientists require more salary to work in Germany than in Britain. (Alternatively, the best of the national graduates would rather work in a national firm.) That means, other things being equal, it behoves national firms for their national universities to train as many chemists as they want to hire.

There are also national markets for dyestuffs. The British market is significantly larger than the German market throughout the period under analysis. Other things being equal, it behoves British firms to sell in the British market, and German firms to sell in the German market. However, the advantage of national firms can be offset if a foreign firm offers a richer menu of dyestuffs. Under our specification, if a foreign firm does more R&D than a national firm, it can usurp the latter's market, at least partially.

Several key dynamic processes and factors influence our story. To a first approximation, the profits of a firm, gross of its R&D spending, are an increasing function of its level of technology, defined in terms of the quality of the dyestuffs it offers and its volume of sales. This first approximation, however, must be modified by two factors. One is that the profits of a firm that does R&D depend on whether the chemists it hires are national or not. The other is that, for a given level of the other variables, British firms earn somewhat more, reflecting their advantaged location regarding the market.

R&D is funded out of profits, but not all firms invest in R&D. Firms can spend nothing on R&D (as do old-style firms), or they can invest a fraction of their profits in R&D (as do new-style firms). Initially, all firms are 'profitable enough' to be able to afford a small-scale R&D facility. Some (the new-style firms) choose to do so, and others choose not to. If the profits of a new-style firm grow, it spends more on R&D.

Given the availability of the new technology, it is profitable to invest in R&D and, given the competition from new-style firms, firms that don't invest in R&D lose money. This is so both in Germany and in Britain. In both countries a certain fraction of firms begin investing in R&D when the new technology arrives. These profitable firms expand, and the unprofitable ones contract. As firms that do R&D expand, their demand for trained chemists grows too. National firms hire nationally trained chemists first, and then (at higher cost) foreign trained chemists.

The supply of chemists provided to industry by universities is a function of the funding those universities receive from government. For a variety of reasons the supply of German chemists initially is much greater than the supply of British chemists. This initial cost advantage to German firms that do R&D is sufficient to compensate for their disadvantage regarding the location of the product market. Moreover, over time, the political strength of the national industry association, and the amount of money it can induce government to provide to national universities, is proportional to the size of the part of the national industry that undertakes organized research.

Start the dynamics just before the advent of the new scientific understanding that leads to the new technique for creating innovative dyestuffs.

There are more (and bigger) British firms than German firms at this initial stage, reflecting Britain's closeness to a large part of the market. No firm has as yet an industrial research laboratory. The supply of chemists being trained at German universities is ample to meet the limited demands of German firms, and British firms, for chemists.

Now, along comes the new scientific technique for creating new dyestuffs. Some British firms and some German firms start doing industrial R&D on a small scale. They do well and grow. The demand for university-trained chemists grows. Since most of the existing supply of chemists, and the augmentations to that supply, are German-trained, German firms are able to hire them at a lower price than can British firms. The German firms that invest in R&D do well on average, relative to British firms and their German competitors who did not invest in R&D. They grow, and as they do their R&D grows. The effectiveness of German university lobbying for government support for training chemists increases as the German industry grows. You can run out the rest of the scenario.

Promise and challenges

I believe the conception of institutions as defining or shaping standard social technologies is coherent, broad enough to square with the concepts of institutions proposed by other scholars, and particularly well suited to be brought into evolutionary economic growth theory. The concept of social technologies as standard ways of doing things when the doing involves interactions among different people or organizations pairs up nicely with the conception of physical technologies as recipe-like, but which is mute regarding the organization of labour.

In my view, the advance of physical technologies continues to play the lead role in the process of economic growth. In the example of the rise of mass production, social technologies enter the story in terms of how they enable physical technologies to be implemented. In the case of the rise of the industrial R&D laboratory, new social technologies are needed to support activities that create new physical technologies. Perhaps a useful way of looking at this obvious interdependence is to posit, or recognize, that physical and social technologies co-evolve. And this co-evolutionary process is the driving force behind economic growth.

Notes

1. This essay lays out a way to bring economic institutions into evolutionary economic growth theory that I think has considerable promise. Those of us developing evolutionary growth theory have known for a long time that this needed to be done. The question was how to do it. I have been working on this question for a number of years, and now think I see a natural way to do it. This essay is basically a report on that work.
2. For a discussion see Nelson (1998).

References

Basalla, G. (1988) *The Evolution of Technology*. Cambridge: Cambridge University Press.

Bijker, W. (1995) *Of Bicycles, Bakelites, and Bulbs*. Cambridge: Cambridge University Press.

Chandler, A. (1962) *Strategy and Structure: Chapters in the History of the Industrial Enterprise*. Cambridge: MIT Press.

——— (1977) *The Visible Hand: The Managerial Revolution in American Business*. Cambridge: Harvard University Press.

Coase, R. (1937) 'The nature of the firm', *Economica*, 4 (November): 386–405.

——— (1960) 'The problem of social cost', *Journal of Law and Economics*, 3 (October): 1–44.

Commons, J. R. (1924) *Legal Foundations of Capitalism*. New York: Macmillan.

——— (1934) *Institutional Economics*. Madison: University of Wisconsin Press.

Constant, E. (1980) *The Origins of the Turbojet Revolution*. Baltimore: Johns Hopkins.

Freeman, C. (1982) *The Economics of Industrial Innovation*. London: Pinter.

Hayek, F. (1967) *Studies in Philosophy, Politics, and Economics*. London: Routledge and Kegan Paul.

——— (1973) *Law, Legislation, and Liberty, Volume I: Rules and Order*. London: Routledge and Kegan Paul.

Hodgson, G. (1988) *Economics and Institutions*. Cambridge: Polity Press.

——— (1993) *Economics and Evolution: Bringing Life Back Into Economics*. Cambridge: Polity Press.

Landes, D. (1970) *The Unbound Prometheus*. Cambridge: Cambridge University Press.

Langlois, R. (1989) 'What was wrong with the old institutional economics (and what is still wrong with the new)?', *Review of Political Economy*, 1 (3): 270–298.

Lundvall, B. A. (1992) *National Systems of Innovation*. London: Pinter.

Metcalfe, J. S. (1998) *Evolutionary Economics and Creative Destruction*. London: Routledge.

Mokyr, J. (1990) *The Lever of Riches*. New York: Oxford University Press.

Mowery, D. and R. R. Nelson (1999) *The Sources of Industrial Leadership*. New York: Cambridge.

Murmann, P. (1998) *Knowledge and Competitive Advantage in the Synthetic Dye Industry: 1850–1914.* Thesis. New York: Columbia University Business School.
Nelson, R. R. (1993) *National Innovation Systems: A Comparative Analysis.* New York: Oxford University Press.
——— (1998) 'The agenda for growth theory: A different point of view'. *Cambridge Journal of Economics*, 22 (4): 497–520.
Nelson, R. R. and B. Sampat (2001) 'Making sense of institutions as a factor shaping economic performance'. *Journal of Economic Behavior and Organization*, 44 (1): 31–54.
Nelson, R. R. and S. G. Winter (1982) *An Evolutionary Theory of Economic Change.* Cambridge: Harvard University Press.
North, D. (1990). *Institutions, Institutional Change, and Economic Performance.* Cambridge: Cambridge University Press.
North, D. and J. Wallis (1994) 'Integrating institutional change and technological change in economic history: A transaction cost approach', *Journal of Institutional and Theoretical Economics*, 150 (4): 609–624.
Petroski, H. (1992) *The Evolution of Useful Things.* New York: Knopf.
Rosenberg, N. (1976) *Perspectives on Technology.* Cambridge: Cambridge University Press.
Saviotti, S. (1996) *Technological Evolution, Variety, and the Economy.* Cheltenham: Edward Elgar.
Schumpeter, J. (1942). *Capitalism, Socialism, and Democracy.* New York: Harper & Row.
Silverberg, G., G. Dosi and L. Orsenigo (1988) 'Innovation, diversity, and diffusion: A self organizing model. *Economic Journal*, 98 (393): 1032–1054.
Smith, A. *The Wealth of Nations* (1937 [1776]). New York: The Modern Library.
Soete, L. and R. Turner (1984) 'Technology diffusion and the rate of technical change'. *Economic Journal*, 94 (375) November: 612–623.
Veblen, T. (1915). *Imperial Germany and the Industrial Revolution.* New York: Macmillan.
——— (1919). *The Place of Science in Modern Civilization and Other Essays.* New York: Huebsch.
Vincenti, W. (1990) *What Engineers Know and How They Know It.* Baltimore: Johns Hopkins.
Williamson, O. (1985) *The Economic Institutions of Capitalism.* New York: Free Press.

3 Towards an Evolutionary Economic Approach to Sustainable Development

J. B. (Hans) Opschoor

Introduction

Much of development economics has centred on structural change, driven by capital accumulation, in developing economies (Martin 1991b). Key themes have included patterns of economic change, policy interventions to accelerate these changes and the role of international flows of commodities and finance in structural change. The importance of technology and innovation are also recognized, and this has since the early days provided an openness to 'evolutionary' or 'neo-Schumpeterian' thinking (ibid.: 27, 51). This essay is part of an attempt to identify how and to what extent development economics might benefit from adopting the evolutionary perspective.

Darwin, presenting the concept of evolution explicitly as a generalization of "the doctrine of Malthus" (Darwin 1859: 117), placed at least some of the origins of evolutionary thinking firmly within early political economics.[1] Evolutionary thinking is now again coming home to economics through an interesting new approach to economic phenomena called *evolutionary economics*[2] which, incidentally, has nothing to do with (neo-)Malthusianism. In addition to what Nelson offers on this theme (see Nelson, this volume), I explore what this perspective might yield when applied to the domain of the interactions between societies and their natural environments.

The relevance of such an application is not at all self-evident. Kurt Martin, a believer in the need for a 'new development economics' that takes on board new aspects and dimensions, has only one rather cryptic passage dealing with environmental issues:

> Ecology... has been neglected. An Indian author once said that it is poverty that is our real pollution (Martin 1991b: 52).

Persaud, in a less cryptic statement, rightly observes that in the literature on development economics "the environment does not get the attention that it deserves" (1997: 74).

V. FitzGerald (ed.), *Social Institutions and Economic Development,* 23–53.

As my tribute to Kurt Martin, I wish to look at the analysis of ecological aspects of development from an evolutionary perspective – not necessarily as a believer in that paradigm, but as one interested in seeing the extent to which an evolutionary perspective might enhance the profundity and relevance of development economics and, possibly, make development more sustainable.[3] In so doing, I incorporate concerns and views that go back to early economists (the Physiocrats, Smith, Malthus, Ricardo, Mill and Marx) as well as notions from younger academic fields such as environmental and 'ecological' economics.

The wider research agenda into which these issues fit deals with a range of questions:

- How do environmental constraints affect patterns of socio-economic development?
- How does economic development affect environmental constraints?
- How do economic institutions, policies and technologies affect these constraints?
- How do economic institutions, policies and technologies evolve and how does their evolution affect environmental constraints?
- Can this process of evolution be directed or accelerated so as to effectively push away environmental constraints?

Ecology and evolution from an economic perspective

If, as Aristotle asserted, humans are *zoon politikon* ('political animals'), then economic analysis may find inspiration in approaches developed in the life sciences, if it wishes to develop well founded theories about societal change. Marshall's view of economics as "a branch of biology broadly interpreted" (1920: 637) suggests as much. An evolutionary perspective might fit such a view. This section reviews in general terms what an evolutionary approach might contribute to the analysis of economic development and change. It starts from an exploration of the possible merits of regarding the biological concept of evolution as a metaphoric or heuristic device for understanding forces driving economic development and change.

In our exploration, however, we must be careful to avoid two traps. First, we should not mistake analogy or similarity for causality, by extending the area of validity of biological principles or features too easily to the cultural domain. In particular, some views by biologists on evolutionary processes might suggest analogies in the social sciences in the form of socio-biological views that could imply far too reductionist a position. Second, we should not reduce evolution to processes driven by competition. The phrase *evolution* originally referred to the processes of 'mutation' and 'natural selection' through a search for subsistence and livelihood. The 'search for livelihood' element soon narrowed to mean 'struggle for existence' or competition, whereas nowadays many (including many economists) would like to explicitly recognize the role of mechanisms such as cooperation, in cultural as well as in natural systems.

Evolution and dynamics

Evolution in its most general sense can be defined as a pattern of change in the structure of the universe, or parts thereof, in space and time (cf. Boulding 1981: 9). Within this broad concept Boulding (op. cit.) distinguishes several evolutionary patterns: physical evolution, biological evolution and societal evolution. Here we are concerned only with the last.

To some degree the various patterns of evolution can be analysed and described by using a single set of phrases largely taken from biology: *species*, *populations* and *ecosystems*. A *species* is a set of individuals conforming to a common definition (either organisms or abiotic objects including goods and services; see Boulding 1981). A *population* is the number of individuals in a species. *Ecosystems* are defined as the organic communities of biotic species and their relationships with each other and with a given abiotic environment. An ecosystem, seen from the perspective of a given individual or species, provides a 'niche' for that species, the equilibrium population associated with any given set of parameters describing the ecosystem as an 'environment' for that individual or species. The species composition, in terms of types of species as well as their numbers, reflects the possibilities that are embedded in the abiotic conditions and the available (bio)diversity.

If the system is not in equilibrium state, endogenous processes may give rise to changes in the populations of species. The same may result from changes in the biotic or abiotic conditions due to exogenous shocks. Ecological 'dynamics' or succession is defined as a process of structural adaptation in terms of species composition given a certain information base (as laid down genetically) and a set of fundamental environmental characteristics (such as climate and geological substrate).[4] One must clearly distinguish the notion of succession from that of evolution. The difference is that in the process of evolution there is innovation in the sense that new species may emerge.

Evolution has to do with the generation, selection, storage or preservation and transmission of information (Ramstad 1994: 70). Biological evolution does these, respectively, via mutation, competition, genetic fixation and reproduction. Darwin's evolutionary theory revolves around the notion of natural selection. For natural selection to produce evolution, three conditions must be met: there must be variation, there must be heredity and there must be competition or scarcity.[5] Neo-Darwinism combines the Darwinian notion of natural selection with (post-)Mendelian genetics into a view of biological evolution as a pattern of gradualist change as the result of random changes in the genetic code filtered by natural selection. In this synthesis the primary mechanisms responsible for variation are genetic recombination and genetic mutation.

New concepts and theories in biology have cast doubts on the generality of the neo-Darwinian synthesis. Moreover, each of these has interesting analogues in the understanding of social evolution:

- Evolution may not be gradual; jumps may occur ('punctuated evolution').
- Cooperation is an assumed mechanism of social relations in addition to competition.
- There may be forms of transfer of acquired information onto the genetic structure making for a Lamarckian-type dissemination of knowledge.[6]
- A broader view is emerging on selection whereby selection is a process through which over time and through marginal changes, a species 'manages to' adapt to a change in the environment. A wider term that may be useful is sorting (Vrba and Gould as quoted in Van den Bergh and Gowdy 1998). Sorting is differential survival beyond mere fitness-related aspects (such as being in the right place) and may cover branching of species, genetic drift and other circumstances.

Economic evolution

Development economics, and economics in general, is concerned with patterns of change in the structure of an economy and the institutions affecting it. Economics assumes scarcity and looks at ways to improve the resource systems serving the needs of the human species, where success may be measured in some unit of welfare or quality of life and – sometimes – its sustainability. Economics is concerned with efficiency and ways to increase that by changes in technology, the division of labour, trade, growth, etc. Some dynamics here seem comparable to the process of ecological succession: improvement or advance towards a situation of maximum efficiency given an informational base (in the form of technologies and preferences) and a set of basic conditions (in the form of initial resource endowments and institutions). When these 'givens' change, economic evolution may be said to occur.

Marshall (1920) defined 'economic evolution' as the development or 'forward movement', typically gradual in nature, of economic structures through changes in habits, techniques and organizational structures.[7] Economic evolution has also been defined in more functional terms, as a process of cumulative change in the way human provision is effected (Ramstad 1994). The essence of the evolutionary approach in economics lies, first, in its focus on the idea of variation in institutions, technologies and products and, second, in its notion of selection between the options that that variation creates.[8] That is, it is "endogenously generated innovation" (Biervert and Held 1992b). Evolutionary economics thus relates to information on existing variation and the generation and selection of innovation; of new variation that is encodable and transmittable to a next generation. Some of these notions are elaborated on below.

Innovation. Biological purists regard evolution as the result of variation at the genetic level, with the gene the carrier of information and innovation. In cultural evolution no such obvious basic unit exists. Some see institutions and values as the units that evolve (see below). Others (e.g. Boulding 1970, 1981) have designs in mind: product designs, technologies or blueprints. Yet others think of information and innovation in terms of social and technological arrangements. Hayek saw individual economic agents as basic units (Vromen 1997), and some regard products, transactions or institutions ('routines') as the basic elements for study in evolutionary economics (e.g. Nelson).

Whatever the carrier, in cultural systems and economies, variation is not restricted to (equivalents of) mutation and reproduction. Anticipation and learning are possible and, what is more, these processes may lead to deliberate innovation in the form of changing institutions, designs and technologies. That is, evolution is not just the result of blind selection of random forces; it can be goals-oriented, deliberate and cumulative. It is Lamarckian rather than neo-Darwinian. For instance, Nelson and Winter (1982) analyse the process by which traits of organizations are transmitted through time. They take 'rules' or 'routines' to be the economic counterparts of genes.[9] Routines are the replicators through which successful new knowledge or know-how is transmitted. Nelson and Winter assume that changes in routines are goals-directed (i.e. intended). They state explicitly that their theory is "unabashedly Lamarckian: it accepts the 'inheritance' of acquired characteristics as well as variation induced by constraint" (ibid.: 11). They see selection as carried out through competition and profit-seeking.

Finally, in cultural systems innovation may start at the macro level. Whatever the cultural analogue to the gene, creativity and innovation are operative at levels other than the micro level. For instance, that at the structural or organizational (macro-) system level is another. Clark et al. (1995) speak of the co-evolution of both levels as the typical concern in social systems evolution.

Selection. There is debate about the mechanisms of selection in social evolution. Some hold that the analogy with natural selection points to competition as the obvious – and exclusive – social mechanism based on rationally calculating individual organisms whereby only the fittest survive. The market would be a perfect ('natural' as Adam Smith called it) organizational setting for that mechanism to operate. Competition is a key mechanism of selection and of more or less 'creative destruction' (Schumpeter) clearing the way for change. We could interpret the more dynamic forms of neoclassical economics, including neo-institutional economics, as a theory of economic evolution based on competitive selection through markets analogous to the neo-Darwinian, exclusively genetic-mutations-based views on biological evolution.

Others argue on empirical grounds that cooperation is another mechanism operative in social organization, so that radical social Darwinism should be discarded. There are several approaches in addition to the theory of selective competition through markets (Gowdy 1999). These include behavioural approaches (goals-seeking, satisficing, adaptation and learning; e.g. the theories of Nelson and Winter), macro models (e.g. Schumpeter's approach), and in ecological economics we find evolutionary work emerging in the form of (bio)physics-based models (such as those by Faber and Proops). In the view of the 'institutionalists' the market is a specific social institution, amongst others reflecting a more cooperative or collective approach to action and transaction (see e.g. Ostrom 1990).

Evolutionary economics

Evolutionary economics has been defined as the study of the part of cultural evolution that deals with the generation, storage, selection and transmission of alternative ways of producing things and with the allocation of that which is produced (Costanza et al. 1993). It has typically dealt with issues related to such changes as in technology, institutions and means of payment. This section delves into some milestones along its path, paying particular attention to theories on the evolution of institutions and on development in general.

Approaches and developments in evolutionary economics. Hodgson (1997) created a taxonomy of evolutionary economics making use of four criteria: (i) the role of 'novelty', (ii) the adoption of reductionism and (iii) of gradualism and (iv) the use of biological metaphors. In his view, the core of evolutionary economics is the subset of 'novelty embracing, anti reductionist' approaches, which would include Veblen, Boulding, Nelson, Winter and Commons. He refers to this wing as the 'institutional wing' of evolutionary economics.[10]

Evolutionary economics differs from mainstream, neoclassical economics in four main ways (see e.g. Costanza et al. 1993, Van den Bergh and Gowdy 1998). First, it does not focus on optimization. Rather, it takes a more behavioural approach in which learning and adaptation are key elements and in which the possibility of numbers of equilibria arises much more easily than that of a single one. Second, evolutionary economics sees increasing returns and positive feedbacks as significant phenomena. Third, it takes contextuality, path-dependency and lock-in as relevant. Fourth, it sees human economic agents as capable of learning and of rapidly disseminating the results of that learning through cultural mechanisms rather than biological ones, as capable to some degree of foresight or anticipation, and as capable of designing new technologies and institutions.

Within mainstream economics, evolutionary approaches are typically traced back to Schumpeter's theories of innovation and qualitative change in firms and markets. Boulding (1970, 1981) redefined production and development in evolutionary and ecological terminology (see above). Societal evolution and innovation then appear as Lamarckian processes rather than as Darwinian 'blind' ones. This explains the difference in pace between natural evolution and societal evolution. Evolutionary theory in general and particularly evolutionary economics currently lean more on models that involve notions such as co-evolution, cooperation and punctuated equilibrium.

Clark et al. (1995) speak of the co-evolution of both levels as the typical concern in systems evolution. From that insight another follows: If both levels within a system (i.e. the micro and the macro; the system elements and the system as a whole) co-evolve, then conditions at any point in time have a particular history, dependent on past innovation and adaptation. Long-term success of any group of agents within the system is, say Clark et al. (1995), possibly more related to the ability to deal with uncertainty and change, than to the ability to identify optima given any set of conditions surrounding that group of agents.

An entire branch of evolutionary theory uses game theory to explore to what extent and under what conditions mechanisms of cooperation would override competitive tendencies, even in a world full of 'selfish geniuses' in the form of exclusively self-interested and arithmetically highly endowed specimen of *homo economicus*. It does appear that in a significant set of circumstances cooperation would be rational (i.e. efficient) as a survival strategy (e.g. Axelrod 1984) and honesty/trust would pay off as well (Van der Klundert 1999). This may be regarded as one example of how values might emerge through an evolutionary process.

Evolution of institutions. Institutional analysis studies institutions, habits, rules, etc., as well as their evolution. In traditional institutional economics not individuals but institutions are the basis of analysis. Institutional economics sees regularities at the systemic or macro level as being reinforced by feedback coming from, and acting upon micro actors. In fact, one assumes a substantial amount of regularity in behaviour based on systemic inertia, imitation, lock-in and cumulative causation.

Veblen was the first to explore economic and institutional evolution along Darwinian lines (Hodgson 1998). His 'old' institutionalism offers a radically different perspective (from that in mainstream economics) on the nature of human agency, based on the concept of habit. That notion, interestingly enough, returned in Marshall's definition of economic evolution (see above).

The so-called 'new institutionalist' approach explains the emergence of new institutions by reference to a model of rational individual behaviour. The explanatory movement is therefore from individuals to institutions. Individuals and their preferences are the given. In old institutionalism the individual was not a given: he or she is both a consumer and a producer of circumstance and environment; the reasoning there flows from the institution to the individual (Hodgson 1998).

Habit and routine are the links between institutional economics and evolutionary economics. They adapt or 'mutate' as agents try improvements that are rewarding from their own perspectives. Learning is a core concept, including the development of new representations of the environment in which agents operate. North's starting point is an individualistic choice-theoretic approach, but the focus is on human cooperation, especially cooperation that enables the capture of gains from trade and labour division. He believes incentives to be the underlying determinants of economic performance (North 1990: 135), but also that institutions may create more hospitable environments for cooperative solutions to complex exchange, with (more) economic growth as the bonus; or "when it is costly to exchange, institutions matter" (North 1991: 12). According to North (1990) institutional change can be driven by at least two forces: firstly increasing returns and secondly imperfect markets and the level of transaction costs. These factors may in fact strengthen the path-dependency of institutional change (or any economic change, for that matter; ibid.: 112).

From an institutional economics perspective, a theory of process, evolution and learning is needed starting from institutions to explain their changing, rather than from an institution-free state of nature with only individuals. Evolutionary thinking attempts to provide that and, at the same time, wants to break away from equilibrium-thinking, comparative statics. For instance Nelson developed theories on how technological innovation and institutional change (as 'social technologies') contribute to enhancing economic performance in the form of economic growth (Nelson and Winter 1982; Nelson 1994; Nelson and Sanpath, unpubl.; Nelson, this volume).

Economic development. Economic development is a complicated process within a complex system of social and society–nature interactions. Most development economics restricts itself to the subdomain of the economic aspects of societal interactions. The interactions with the natural environment are brought in as an explicit point of departure in the remaining sections of this essay.

Old theories of development like Rostov's theory of stages of development describe a sequence of stages through which economies evolve with time and economic growth. One question we may ask is whether these stages exist, how to discern them and how to monitor the implicit process of evolution. Another has to do with dynamic issues: Are these structural changes income- or output-dependent only, or do context (including ecological conditions), history and culture play a role as well. In a less retrospective, descriptive and analytical mode (i.e. in seeking strategies of development and methods of implementation) one may also be concerned with the costs of transition and the benefits of induced development.

In a perspective of systems dynamics and systems evolution, when economies are developing or even evolving, they are by definition in a non-steady state. Exogenous or random shocks may push systems-to-be away from equilibrium. Typical of systems away from equilibrium is that there may be self-reinforcing mechanisms (positive feedbacks) possessing asymptotic states or emergent structures towards which the systems are pushed. Thus, structures are 'selected' that the system eventually locks into (Anderson, Arrow and Pines 1988). As opposed to this, normal economics typically is about systems with negative feedbacks and diminishing returns.

Nelson deals with economic evolution – in fact, with the 'co-evolution' of technologies and institutions – and the contribution thereof to the process of economic growth and development. Growth and development reflect the way in which available factors of production, particularly capital and labour, are being utilized, given their basic scarcity. Growth theory in general is evolving to include some of the realities that institutional and evolutionary economics grapple with in their development-oriented modes (Ruttan 1998). Recent endogenous growth theories add positive externalities, learning and anticipation, accumulation of knowledge and even 'ideas' to the list of factors to consider when trying to explain economic growth. Still, Ruttan remains critical of the new growth theorists' claim that their work replaces old growth theory as well as development economics. He feels that there are many factors that the new growth literature still has difficulties handling – if it has tried at all. Among these are structural transformation, natural resources constraints and institutional change. With the latter two items we are at the heart of the concerns of this essay.

Some intermediate conclusions

This section has indicated that evolutionary approaches to economic development in the form of structural changes in patterns of production, technologies and institutions may be quite useful in that they draw attention to fundamental mechanisms in the generation, storage, selection and transmission of alternatives in production processes. Yet some fundamental distinctions may exist between processes of biological evolution and cultural ones (of which economic evolution is a part). These cast doubt on simple generalizations from phenomena and regularities in biological evolution to cultural evolution (see Boulding 1970, 1981; see also Van den Bergh and Gowdy 1998).

Economic or societal evolution is much faster than biological evolution. This has to do with learning and the capability to transmit knowledge that allows societies to evolve in a Lamarckian way. The (neo)Darwinian model of evolution through genetic mutation is a meagre one compared with modes of change that seem relevant in economics or in societal processes generally. Also, in biology the gene is the preferred unit of selection (although some speak of macroevolution and higher level sorting processes). There is no unanimity about the gene's cultural or economic equivalent.

In the context of our wider interest in the relationships between development and the environment some potentially useful notions in evolutionary economics have emerged:

- 'bounded rationality' or a constrained knowledge of the set of options, and rapidly increasing costs of information in overcoming these constraints;
- the notion that there is not a general 'best', but that contextuality poses constraints (cf. the debates in development studies), which in evolutionary jargon can be described as related to lock-in and irreversibility;
- structural change, irreversibility, path dependence and lock-in;
- notions related to learning and adaptation;
- the idea that change can be abrupt and non-gradual (though this seems well entrenched at least in earlier evolutionary economics).

Environment–economy interactions and ecological economics

Most conceptual models of the economic process regard it as a set of circular flows, real as well as monetary. The models are represented as a closed system with respect to material and energy flows: they are there, given and eternal. Economists have thus perceived the economic process as a kind of economic *perpetuum mobile*, which any natural scientist will immediately recognize as impossible. Environmental economic theories take environmental problems on board, essentially as constraints on production, consumption or welfare. These problems are either exogenous (they come from outside the domain of the economic process) or they are spinoffs from activities by other economic agents ('externalities').

But such an approach does injustice to the dynamics of the interactions between economic processes and biospheric ones. Ecological economics adds at least three aspects to the frame of analysis of traditional development economics. First, environmental capital or natural resources need to come in as forms of capital, to conceptually capture the essence of production (as is the case in physiocrat as well as classical economics in the form of 'land'). Second, environmental quality may be a contributor to human welfare. Third, economic growth through its levels and patterns may eventually erode the overall capital base and adversely affect levels of human welfare.

The appropriate view of the system is thus not that of a closed *perpetuum mobile*, but rather that of an energetically open system: the biosphere and the economy within it. Economic processes occur within a physical environment (and are subservient to the laws that prevail there, such as those on the conservation of energy and matter and their corollaries on entropy). They are in open interaction when it comes to matter and energy. Being part of a wider system makes the economic one susceptible to direction, historicity or – more generally – contextuality. Development economics should therefore incorporate those environmental considerations that are relevant to the economic process, *including* appropriate aspects (and parts) of the natural environment (see e.g. Opschoor 1996).

In an expanded system, then, environmental problems no longer appear to be largely exogenous but more often take on an endogenous or semi-endogenous nature. This expanded system is multi-modular (see e.g. Duncan 1959, Wilkinson 1973, Opschoor 1996). Moreover, parts of it – if not the system as a whole as well – show chaotic features that make understanding and predictability even more remote. Costanza et al. (1993) list additional features adding to the complexity: the multi-layer or hierarchical aspects of the expanded system, its non-continuities and its state far from equilibrium.

Since many environmental issues call for analysis in the (very) long term, evolutionary processes – not so much in nature but in society – may indeed come into play. This has to do in the first place with endogenous structural change within society and its institutions. But evolutionary notions may come in as well because of the need to adapt to a changing natural environment, whatever the roots of those changes (i.e. whether natural or socially induced).[11]

One must view environment–economy interactions as changes in the 'co-evolving' ecological and economic systems (e.g. Norgaard 1984, Leipert 1992) and as changes at the micro level as well as the macro. The latter type of co-evolution implies that agents are much more the makers or producers of their own environment than mere victims of fate and inevitability. Combined, both types of co-evolution lead us to conclude that the natural environment needs to be incorporated into our models of the systems within which economic agents function; we need to improve our capability to understand and model all relevant processes including ecological ones, at least to the extent necessary to be able to deal with change in the overall system and to reduce uncertainty there. This allows us to cope far better with variabilities in our environment and to generate appropriate responses. As Clarke et al. (1965) remark, we need to "endogenize the environment" (or rather, the relevant parts of it) in order to more effectively anticipate societally relevant change at all levels.

Academics try to chart parts of this expanded system and explore its development or evolution to some degree. They may do so more urgently if they receive signals that the interactions between economic subsystems and the rest may give rise to potentially unpleasant surprises (e.g. climate change). They may wish to anticipate these or develop a capacity to anticipate, which, as we shall see, is attempted through scenario analyses and integrated assessment in environmental sciences. In an *evolutionary* perspective this makes sense, as it is a precondition for learning about new needs to adapt to biophysical realities by changing patterns and levels of economic activities, the technologies used in them or the institutions governing development.

Evolutionary dimensions in ecologically sensitive development economics

Nelson and Winter observe a need for more insight into "the conditions for survival in an evolutionary struggle in a changing environment... in which... routines are responsive to environmental variables" (1982: 408). This holds for firms in a changing *economic* environment. I have argued that this is also true when we consider economies in a much broader sense, including their links to the *physical environment*, and when we analyse not only the dynamics of these links but also the role of old and new institutions in terms of changes in the natural environment. This idea is well known in ecological economics and related disciplines like human ecology and geography.

Societal evolution and environment: Historical perspectives

Co-evolution of society–nature interactions through development, industrial modernization and even ecological restructuring is a key feature of current development patterns. As such, it has to enter into a new development economics.

In analysing the economy in the context of broadened systems, Duncan (1959) kicked off by discerning within the expanded system a series of modules (or subsystems) for technology, culture, organization, population and environment. There were dynamics endogenous to each of these modules (e.g. population dynamics) as well as interactions between them (e.g. environmental degradation driven by population growth). Wilkinson (1973) developed these notions into a historical, 'dynamic' ecological model of economic development, which views cultural, organizational and technological change as driven by 'poverty' in terms of resource scarcities resulting from population growth. The model seems to hold for pre-modern societal development, but it does not seem to capture developments since the industrial revolution. Others would argue that the model's Malthusian pessimism does not really hold for either past or current patterns of change in developing societies (e.g. Boserup 1975).

In a recent review, Boserup uses the notion of structures (where Duncan had 'modules') as entities with a certain stability but susceptible at least to influences from other such structures. Boserup (1996: 505ff) analyses different development theories, such as those of Smith, Malthus, Ricardo, Marx, Weber, and the neo-Malthusianists, to show these authors' differing views on development in terms of the assumed links between these structures. This provides a point of departure for an 'evolutionary ecological' approach to development, in which new structures, relationships, institutions and technologies might emerge: the process of social change within the overall system is seen as turning from a 'succession'-type process into an evolutionary one.

Institutional evolution and environment

Institutions affect economic performance including growth. In the short run and in the absence of built-in ecological concerns this might directly or indirectly give rise to aggravated environmental pressures from economic activities onto the physical environment, and even to an enlarged risk of ecological unsustainability. In the long run environmental impacts may trigger new institutions in relation to the perceived urgency of the unsustainability. Depending on the accuracy of that perception, the institutional innovations may be insufficient to curb the latter. This points out the need for analytical and empirical research on the factual evolution of the impact of institutions and policies in response to environmental change. It also points in the direction of prospective, design-oriented research on such institutions, to make them more effective in achieving sustainability. Three specific aspects of such research can be mentioned:

- Lack of information and knowledge on the environmental repercussions of economic activities sometimes entails costly processes of approximating these repercussions. Or, alternatively, an information deficit may lead society to accept decisions based on imperfect information. Decision-making in a procedurally reasonable way in the light of available knowledge and means of computation is one way to deal with this (North 1991, from Simon 1986).

- There may be imbalances in prevailing systems of access and property rights affecting the readiness to invest in the ecological sustainability of economic processes at the micro, meso and national levels.
- Poor institutions may make for high uncertainty, short time horizons, cost shifting, etc. Costs are associated with change. These transition costs (as I would call them) may in fact bar institutional reforms that could bring about sustainability.

Norms and values

Generally, values and similar predispositions evolve as institutions in the process of repetitive game-like strategic situations (Van der Klundert 1999). This finding may support some hope that new values such as 'sustainability' might indeed not only emerge, but, under the right conditions, even have a chance of becoming dominant. The urgency of the emerging unsustainability, the realization of a strong anthropogenic component in the causal processes behind unsustainability, a sense of responsibility for future generations and other lifeforms, the view that society and its institutions are socially constructed to some extent and several other notions and conditions may be relevant here. We are reminded of Rawls' approach to justice and an intertemporal extension of it suggesting a maxi/min-type solution on the intertemporal distribution of resources. However, there never will be a true veil of ignorance between the present and the future; there is no real consecutive game with future generations. Evolutionary economics, in contributing to sustainable development, might explore conditions for the emergence of sustainability-related values and institutions in more detail.

The Environmental Kuznets Curve

One prominent issue in debate about sustainable development is the relationship between economic growth and environmental degradation. Relevant here are the evolution of economic production patterns, technologies used, environmental awareness and policies and institutions as per capita incomes increase. I shall review some research on this relationship, from what might be considered an evolutionary perspective on it.

Environmental macroeconomics

The interactions between economic activity and the environment have been metaphorically labelled society's 'industrial metabolism' (Ayres 1994). The aggregate flow of matter and energy involved in this metabolism has been called 'throughput' or the economy's 'scale' in terms of its ecological claims (Daly 1991a). We use the symbol S for it. The study of the economics of this at the macro level has been named environmental macroeconomics (Daly 1991b). Metabolic efficiency is defined as the economic output per unit of throughput. Economic output is often measured by the level of production or income, Y, so metabolic efficiency is measured as Y/S. The inverse of this, S/Y, is the throughput intensity, S. A more efficient metabolism is characterized by lower levels of materials and pollution intensities.

S can be defined as the product of population level P, income per capita y ($= Y/P$) and the throughput intensity S:

$$S_t = P_t \cdot y_t \cdot s_t \; (= s_t \cdot Y_t) \qquad (1)^{12}$$

This equation is a condensation of what happens on the economy–environment interface: how environmental pressure in the form of throughput evolves with income. In simplistic analyses, they are linearly related so that S evolves proportionally with Y: this means that s_t is assumed to be constant. The propositions on which some Club of Rome debates are based have sometimes reflected this assumption. In fact, s_t could be a function of income. The World Bank (1992) in its contribution to the UN Conference on Environment and Development even assumed that s_t might evolve, from low initial values of y_t, so as to make S_t grow at a decreasing rate to, beyond some threshold value, even decline in absolute terms. S_t, in that perspective, would have the shape of an inverted U. That relationship (i.e. S_t as a quadratic, peaked function) is referred to as the 'Environmental Kuznets Curve' (EKC). We shall return to its empirical validity and relevance below.

Taking the first derivative of equation (1) we get

$$dS_t/dt = Y_t \cdot ds_t/dt + s_t \cdot dY_t/dt \qquad (2)$$

For an EKC-shape to emerge, dS_t/dt must at some point be less than 0. So in that segment

$$(-ds_t/dt)/s_t > (dY_t/dt)/Y_t \qquad (3)$$

Thus, the rate of change in the throughput intensity must exceed (in absolute value) the economic growth rate. 'Relative delinking' is the situation where there is delinking (i.e. $ds/dt < 0$) but where the effect of this on S is overtaken by that of economic growth. So, S would increase despite an enhanced metabolic efficiency. 'Absolute delinking' is a situation where equation (3) holds.

The EKC expects relative delinking up to the threshold and absolute delinking beyond that. However, there may be a point beyond which throughput rises again with per capita income. This may be related to diminishing returns to investment in environmentally friendly innovation or even to the existence of technological (thermodynamic) asymptotes to the reduction of throughput intensities, at least along known innovation trajectories. From that point onward, the economy and environmental pressure S will be *relinked* (Opschoor 1990), at least until further breakthroughs in research and development occur, or until a more intensive application of environmental policy checks is implemented. We call this prediction the 'relinking hypothesis'; empirical manifestations of $dS/dt > 0$ in specific areas of environmental concern could be taken as validating it at least in part. Growth optimists believe that absolute delinking can be a permanent phenomenon; growth skepticists are concerned that possible relinking could erode the sustainability of economic development.

Be that as it may, absolute delinking should manifest in the economically more advanced (in terms of per capita income) countries of the world, that is, in the OECD region. Elsewhere, countries are still expected to be in the upswing or linkage phase or, at best, at the start of the downswing. Knowledge about the magnitude of this phenomenon, the turning points, contributions of the various factors and its persistence would be of interest to most countries.

Empirical EKCs

The above reflects where ecological economics was around 1990. Since then, research into the relationship between economic growth and environmental quality has gained relevance due to some overly optimistic interpretations of the work by the World Bank. If the EKC exists – so these interpretations went – then we can just push for economic growth everywhere and sustainability will follow. Much effort and finance has since gone into empirical work on this topic, also by my own group at the Free University (see De Bruyn 1999, De Bruyn and Opschoor 1997, De Bruyn, Van den Bergh and Opschoor 1998). In my inaugural address at the Institute of Social Studies I could, on the basis of that work, report the possibility of the relinking hypothesis based on observations on 20 countries in the North – OECD and (former) COMECON – over some 25 years (Figure 1). Where are we now? What are the links between growth and environment? Do we understand them better? Are they a matter of simple dynamics of the succession type, or is there evolution in this chaos?

Global cross sections. Many studies reveal decreasing material and energy intensities in a range of OECD countries especially during the 1950–80 period (for a review, see De Bruyn and Opschoor 1997). Also, data on pollution in relation to economic development often show reductions of pollution per unit of production or income as income levels rise. They often even show re-

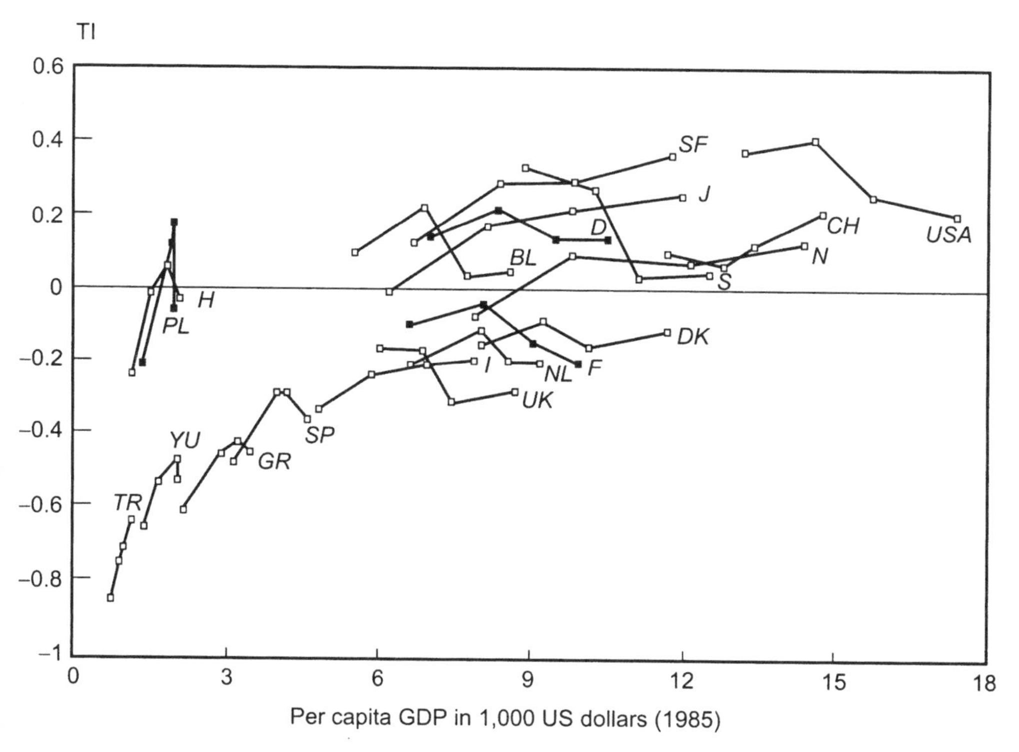

Figure 1 Relinking: Developments of selected countries in levels of TI and per capita GDP, 1966–90

Source: De Bruyn and Opschoor 1996.

ductions in absolute levels of pollution (World Bank 1992, Shafik and Bandyopadhyay 1992, Selden and Song 1994) with varying turning points.[13] Against this trend, Sengupta (1996) investigated CO2 emissions for a sample of 16 countries ranging from India to the United States excluding economies in transition (but including China) over 1971–88. He found a best fit with a polynomial of third degree, suggesting the existence of an N-curve. Similar N-shaped curves have been found for other pollutants and for energy consumption.

To empirically test the delinking hypothesis at the level of total throughput, De Bruyn and Opschoor (1997) applied time-series analysis (1966–94) on 20 OECD and COMECON countries. It appears that the economies considered have indeed delinked but are entering a period of relinking, cf. N-shaped patterns.

Transition economies. In Central European countries over 1970–91, levels of throughput were high in comparison with Western Europe, both in absolute terms and per unit of GDP. As in Western Europe, throughput per unit of production dropped in the 1980s, and the centrally planned economies actually began to environmentally delink before 1989 (Rebergen and Opschoor 1994, Opschoor 2001). This may have been due to attempts by previous regimes to increase efficiency, partly in response to more effective societal concern over environmental issues and partly owing to the emerging changes in patterns of production.

Energy consumption, steel production and several types of emissions (particulates, SO2, NOx and CO2) were analysed for Poland, Hungary, former CSFR, Czechia and former Eastern Germany. All these countries showed relative delinking, at least after the late 1970s, with absolute delinking setting in after 1984. After 1991 these economies showed economic recovery (in terms of rising GDP), but most throughput indicators continued to decline at least initially. If growth rates pick up, the effects of relative delinking should quickly be overtaken by the growth effect and in absolute terms there will be relinking.

Developing and emerging economies. The relationship between environment and growth in *developing countries* has been studied mainly from the perspectives of deforestation and energy use. On deforestation the evidence diverges as to whether or where it follows an EKC-pattern and as to turning point (e.g. Panayotou 1993, Cropper and Griffith 1994, Rock 1995).[14] On energy use the situation appears to be one where energy growth has outstripped GDP growth in most developing countries and may continue to do so (e.g. Bernstein 1993, Sengupta 1996, Gupta and Hall 1996).[15]

In an explorative study reported briefly in Opschoor (2001) we analysed nine countries (two low-income countries, four lower middle-income countries and three upper middle-income countries) with a somewhat arbitrary but at least uniform throughput indicator[16] (period 1970–92). Throughput varied with income level in these countries. Linear regressions of this index with PPP income per capita give positive and significant regression coefficients. In other words, the marginal impact of income on throughput declines as countries develop. Or, throughput per unit of GDP increases in low-income countries but seems to decrease in middle-income countries.

There may thus already be some relative delinking in the middle-income countries, possibly as a result of leap-frogging (their use of newer, cleaner technologies than industrialized countries would or could have done at similar levels of income). Nevertheless, due to ongoing economic growth, the overall throughput levels showed increased environmental deterioration. Throughput intensities (throughput per unit of GDP) were calculated (Table 1) and show a rise in low-income countries and a decline in the lower middle-income countries (and, as expected, much more of a decline in the upper middle-income countries); on the whole they increased. It is also interesting

Table 1 Estimated Throughput Intensities in Selected Developing Countries, 1970–91

	1970	1980	1991
Income level:			
low-income countries	0.104	0.139	0.157
lower middle-income countries	0.131	0.108	0.095
upper middle-income countries	0.132	0.083	0.059
Population size:			
small	0.112	0.107	0.101
large	0.137	0.130	0.135
Trade orientation:			
primary commodity exporters	0.119	0.122	0.124
industrializing	0.125	0.100	0.095

Source: Rebergen as published in Opschoor 2001.

to see the evolution of this indicator in relation to trade orientation: in industrializing developing economies throughput intensities decrease whereas they tend to increase in economies exporting primary commodities.

Economic theory and EKCs: Towards an evolutionary perspective

Empirical studies, then, do often indicate decreases in specific forms of environmental pressure with rising average incomes in developed economies. Delinking is indeed becoming visible, both in economies in transition and in developing economies. However, in some areas and at the global level relinking seems to emerge. In addition, even where EKCs are found, they often imply a reduction of metabolism beyond fairly high average incomes, which, given the current levels and distributions of income and people, would imply that environmental utilization may or will keep growing for at least a number of decades, with subsequent risks of unsustainability. Thus, correlations sometimes found between economic growth and environmental improvement aimed at testing the hypothesis of an inverted-U or parabolic relationship must be used cautiously, as they do not lead to unambiguous results. This is no surprise from the perspective of environmental macroeconomics (see above). Now let us explore these results from a more evolutionary viewpoint.

Economic growth typically leads to rising per capita income levels triggering changes in the structure of demand for products and services via embedded differences in income elasticities. This has given rise to drastic changes in the sectoral composition of national products. Economies have evolved from resource-intensive agricultural and mining stages to more industrial ones and seem to be moving towards a post-industrial, service- and information-oriented stage. It is often presumed that this in itself will lead to a reduction of S/Y. Furthermore, an income-elastic growth in environmental demand (given the preference structure and the knowledge base) may show up in larger budgets, both absolute and relative, for environmental policies. These are some of the changes that would emerge *given the preferences and the level of knowledge and information*. The inherent dynamic is of the 'successional' type.

Additionally, with rising environmental pressure, knowledge of ecological functions may increase and become more widespread, altering hitherto prevailing preference schemes and political priorities. This may accelerate the impact on S/Y from industrial development. Moreover, environmental considerations may trigger technological change where resource scarcities and deteriorating environmental qualities became matters of concern and even urgency. These types of drivers of changes in demand and production (and hence in the associated metabolism) may be regarded as 'evolutionary' rather than successional.

Economic factors put forward to explain delinking include (i) structural change in production patterns, (ii) positive income elasticities for environmental policy, (iii) increased levels of trade as income rises, (iv) changing preferences and priorities due to increasing information and (v) changes in production technologies and product designs. My associate De Bruyn has tried to disentangle the roles of these driving forces behind the evolution of throughput intensities (De Bruyn 1999). Essentially, De Bruyn finds that typical economic factors, such as structural change in the patterns of production and changes in the international division of labour and production, do not play major roles. Environmental policies and their accelerating impacts on innovation and on diffusion of new technologies may be more important. We should note that the relative contributions of endogenous elasticity-related forces, on the one hand, and the infusion of new knowledge into societal perception and preference formation, on the other, cannot yet be separated.

We can go further and study the evolution of throughput intensities in a dynamic setting with a tool of analysis taken from Ormerod (1994) called Connected Scatter Plot analysis (CPS). CPS essentially pictures a phase portrait of a variable by plotting its value in year t against that in year t–1. In systems in equilibrium, such diagrams would show variables to circle around some equilibrium value (referred to as the 'attractor point' in chaos theory). If the system is not in equilibrium the variable would be seen to move through the diagram. If the system were on its way to another equilibrium then the variable might be seen to begin orbiting around another attractor point. De Bruyn has applied this device to see what was happening to two key determinants of throughput: steel consumption intensities and energy intensities in the United States, Great Britain, West Germany and the Netherlands. These are clearly four economies potentially over the Kuznets' hill. Figure 2 shows the results for three of these economies.

In almost all cases there indeed seems to be a downward trend in intensities (energy in the Netherlands being somewhat of an exception) away from an equilibrium in the 1960s down to a yet-to-emerge new (and lower) attractor point (which, in the case of steel consumption seems to be emerging). Over a 30-year period these intensities seem to have dropped or are dropping by some 30% to 50%.

This work seems to corroborate some points made above. The inherent process of increasing resource productivity may result in new, fairly stable levels of resource or throughput intensities even in very advanced and innovating industrial economies. Beyond that, we may indeed expect relinking – unless new great shocks such as the oil crisis of the early 1970s were to trigger a new wave of environment- and resource-saving technological innovation or lifestyle change. Unless such new trajectories are sought and found, we should expect cubic N-curves to exist where quadratic inverted-U's were hoped for in mainstream economic documents.

Conclusions

We have been concerned with economic evolution in relation to changes on the interface between society and the environment. In formulating conclusions, it may be appropriate to take our cue from Malthus' prophecies on the breakdown of life support systems and human population numbers as a result of imbalances between population growth and the possibilities for raising the productivity of natural resources ('land'). About 100 years later, Marshall analysed why Malthus' doomsday had not dawned (Marshall 1920: 149, 150). He pointed to the rapid changes in agricultural productivity and the huge improvements in transport technology as having saved humanity the need to either go into moral constraint or go down with Malthus' positive checks of misery and disease.[17] As we would say now, one particular form of cultural evolution (technological innovation) is recognized as having facilitated the pushing away of a serious constraint on development.

Today, some 30 years after Meadows' and Forrester's 1972 prediction of the environmental limits to growth, social scientists claim that societies' turning to institutional change has saved the day again, with a promise to repeat that far into the new millennium. Have we learned how to use technology – physical or 'social' (to borrow a phrase from Nelson) – to continue to save our collective necks in the face of the constraints posed by rapid social change and a slowly evolving biosphere? If not, can society improve its adaptive performance in the face of these constraints? These are interesting issues from a perspective of sustainable global development.

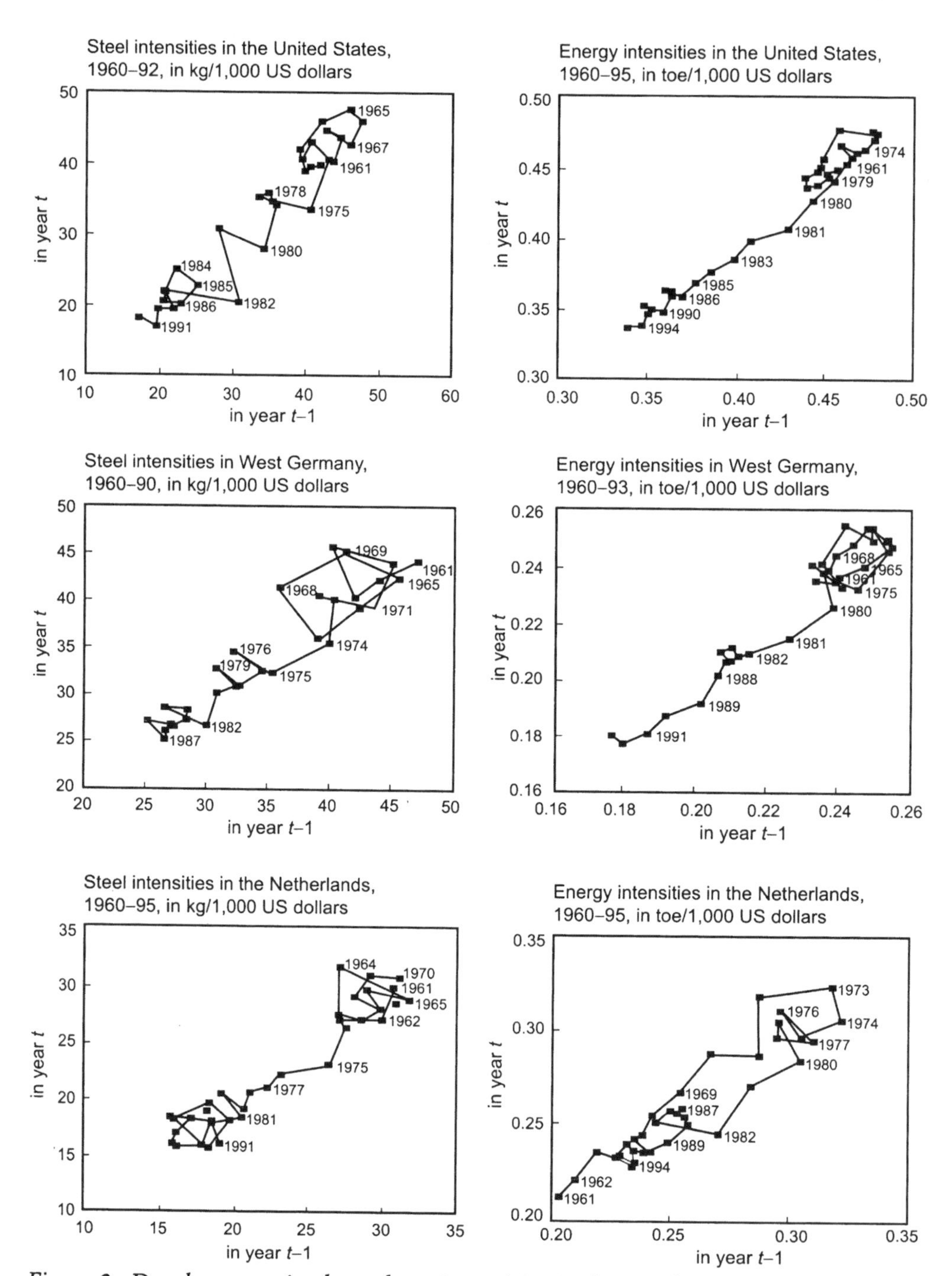

Figure 2 Developments in throughput intensities in three industrialized countries

Note: Amounts are in 1990 US dollars.
Source: De Bruyn (2000).

Addressing these questions seems to require two key changes. First an evolutionary approach needs to be applied to understand changes in institutions, technologies, policies, etc. that are relevant to development and sustainability. Where necessary, such an analysis should further be utilized to design more effective innovation. Indeed there are elements in the society–nature system(s) that warrant an 'evolutionary' perspective that combines dynamics with some form of learning-based innovation – though there are others that do not require such an approach. Moreover, an evolutionary approach cannot be simplistically reduced to a biological, neo-Darwinian view. Evolution in economic and institutional domains may differ from biological evolution:

- It is more clearly driven by macro-phenomena or is more 'punctuated' than its biological counterpart (Gowdy).
- It may be less random and more teleological than its biological equivalent (Boulding; Nelson and Winter).
- There may be more reason to allow for sorting mechanisms beyond mere selection (Van den Bergh and Gowdy) including those based on learning and anticipation.

The second change needed is to advance our analytical and anticipatory capacity to enable society to better understand and predict the dynamics in society–environment interactions and co-evolutionary change in relation to these; that is, a move towards integrated modelling and theory-building including relevant parts of the ecological systems in which economic processes are embedded. The growth literature – old and new – still seems to have difficulty handling phenomena like structural transformation, natural resources constraints and institutional change. As these still pose problems in terms of what Ruttan calls "analytical and modelling tractability" we can expect that modellers will need some time to incorporate these factors adequately, if they are to become successful at all. A more eclectic approach to environment–society interactions and the dynamics these induce seems inevitable, even if some will continue the search for 'first principles' from which to derive it all.

A reformulation of development theory can draw on several insights from ecological and from evolutionary economics, and even combine them into something useful in a development orientation. This essay has discussed some aspects of these approaches. One specific area of review was the field of research on links between economic growth and environmental degradation. There we found that sustained economic growth is not necessarily environmentally sustainable, nor will it automatically become sustainable. Sustainable development is most likely to be the reflection of deliberate environmental and developmental policies and policy-induced technological and institutional innovations. Given the information that is available now on turning-point levels of average income (if there are turning points), with economic and population growth as expected, environmental pressure at the global level will continue to increase far into the future. This point holds especially true for the developing countries and economies in transition. But there is also reason for serious concern among the industrialized countries. The OECD countries may now be entering a phase of relinking. This indicates a need to accelerate policies and find new trajectories of technological innovation to drastically improve metabolic efficiency. If the EKCs of countries are to be 'suppressed' to lower levels, and if turning points are to be pulled forward in time, then technological cooperation and technology transfer might be very relevant.

What can development economics learn form this exercise? Indeed, as Kurt Martin suspected, it has to be broadened. On academic grounds it cannot be a subdiscipline of economics alone, but has to open up to ecological as well as evolutionary aspects.

Development occurs on the basis of activities involving both the natural and the social spheres, and it relates to notions such as learning, anticipation, adaptation, positive externality and feedback. Development takes place in a world that is rather open to the future; in that sense, some of the adaptations and options (technologies, institutions) can be a matter of human design and societal implementation – in a more or less rapid, potentially goals-oriented, cumulative way. Even values and preferences and their evolution can be analysed by, and understood through, evolutionary methods. At the same time, an evolutionary perspective helps us to interpret development as it takes place in terms of contextuality, path-dependency and even in the possibility of 'lock-in' as constraints on processes of social change.

Evolutionary and ecological perspectives might be combined to analyse the robustness and resilience of economies and societies to fluctuations in development, to environmental conditions and to combinations of these two (Costanza et al. 1993). They might also both be needed to understand irreversibility in processes of societal and natural change relevant from a development perspective. The co-evolution of the two systems or levels needs to be better understood to arrive at more appropriate assessments of changes in each. The notion of sustainability may be given more substance and tangibility on that basis as well.

All of the above is possible, and some of it may even be plausible. Yet it seems to me that the main contribution of an evolutionary perspective is not so much its predictive powers nor perhaps even its analytical ones, although it does provide analogies from a more fundamental level of change that may shed light on economic development. From our review it appears that evolutionary economics provides some heuristic tools to indeed obtain more profound insight into what is occurring when there is economic change and development. We are possibly justified in expecting that these more profound insights will give rise to better grounded policy recipes.

Notes

1. This is based essentially on the role of competition in the face of scarcity, referred to by Darwin as the 'struggle for survival'. Marshall traces this type of thinking back to Adam Smith (Marshall 1920: 200). Mill (1848) saw the 'principle of competition' as the one principle giving political economy its scientific basis (see introduction to the Penguin version, Mill 1848: 38) and this was published well before Darwin's *The Origin of Species*.
2. Veblen is said to have argued for an evolutionary economics first (in 1898, as quoted in Hodgson 1997). Although Marshall also used the phrase 'economic evolution' (Marshall 1920: xi) and saw his work in that perspective, he developed a more mechanical and reductionist approach to economics rather than an evolutionary one. Samuelson's work on economic development is the usual reference for economists outside the institutionalist perspective.
3. Sustainable development implies some rather fundamental changes in (i) the way resources are exploited, (ii) patterns and levels of investments, (iii) the direction of technological innovation and (iv) societal institutions. It is common to distinguish at least three dimensions of sustainability: ecological, economic and social/cultural (see e.g. Opschoor 1996).
4. Succession is often believed to lead to more diversity as well as more efficiency in the use of available resources within given ecosystems (e.g. Odum 1971), due to processes of competition and competition-based selection (the so-called 'economy of nature'; Darwin 1859: 116).
5. Darwin himself probably thought that in the end at the human level the basis for selection is the group rather than the individual. He also seems to have queried the universality of self-interest. But according to Watkins (1998) he fell in the reductionist trap of atomistic/individualistic assumptions and accepted that the world is essentially competitive.
6. Lamarck maintained that variation is not random, but purposeful and adaptation-oriented, and that acquired (phenotypic) features can be inherited directly by what later on would be called 'change at the genetic level'. A fundamental process underlying social change is the mechanism of conveyance of information on successful adaptation or development. This operates via 'phenotype reproduction' in production processes and via education rather than via mutation of genotypes and selection (Boulding 1971).

7. Marshall distinguishes a biological approach to economic phenomena from a mechanical one based in the physical sciences. He appears to label the two as economic evolution and economic dynamics, respectively (Marshall 1920: xii). It is interesting to see how Marshall justifies this: economics, like biology and unlike e.g. chemistry, "deals with a matter, of which the inner nature and constitution, as well as the outer form, are constantly changing" (Marshall 1920: 637). In fact, he was so concerned with the need to take a biological, even evolutionary approach, that he gave as motto to his book: "*Natura non facit saltum*", the phrase that has come to typify the neo-Darwinian, gradualist approach to evolution.
8. According to Hodgson (1997: 14), "'evolutionary' processes in economics involve ongoing or periodic novelty and creativity, thus generating and maintaining a variety of institutions, rules, commodities and technologies".
9. 'Routine' is their term for "all regular and predictable behavioural patterns of firms" (Nelson and Winter 1982: 14). An agent's (e.g. a firm's) routines define a list of functions that determine (perhaps stochastically) what it does as a function of...external...and internal...variables" (ibid.: 16).
10. It is interesting to provide Hodgson's qualification of some other relevant 'evolutionary' economists: Schumpeter features as a reductionist novelty embracer, Marshall as a reductionist non-novelty analyst, and Marx as not interested in novelty and non-reductionist.
11. Some of these changes in nature could indeed be related to biological evolution of course, especially to the impacts of selection and sorting on populations and on ecosystem structures, and hence on resources. As said earlier, these forms of evolution are not included explicitly in the analysis presented here.
12. Those familiar with environmental literature will recognize in this the "I = PAT" identity going back to Commoner 1971.
13. Ansuageti et al., in reviewing the turning points, finds emissions into the air to range between PPP$5,000 and PPP$20,000 (and even to PPP$35,000 for CO2). For deforestation they range between PPP$825 and PPP$5,500.
14. Panayotou (1993) finds inverted-U's for deforestation with a turning point of around $825 per capita. Cropper and Griffith (1994) find significant turning points in Africa (PPP$4,760) and Latin America (PPP$5,420) only, not in Asia. Rock (1995) finds differently with a turning point of around PPP$3,500 in Asia.
15. Bernstein (1993) investigated the historical relationship between energy use and economic growth (GDP) for 40 developing countries (1971–87). Energy growth outstripped GDP growth. In terms of energy intensities he finds that developing countries have become less efficient. Sengupta (1996) shows how, for countries such as Brazil, India and Indonesia, the primary energy intensities (PPP) grew 0.6% to 3% annually. He expects that the commercial energy intensity of GDP might grow at levels in the range of 2.25% to 3.5% p.a. (see also Gupta and Hall 1996).

16. The throughput indicator was based on (i) fertilizer use per hectare of crop land, (ii) ratio of area of harvested forest to total area forest, (iii) commercial energy consumption and (iv) number of commercial vehicles per capita. These indices were also aggregated into one index of environmental pressure by taking a nonweighted average of the individual indices (1970 = 100).
17. It is more than ironic to note that Marshall, in a footnote to his paragraph on Malthus (Marshall 1929: 150) predicts that, despite much technical progress, the Malthusian pressures of population growth would exceed biospheric carrying capacities around the year 2100, when he expected world population to rise to the 6-billion level. In fact, Marshall's prediction in terms of time was off by a century. We celebrated the advent to the 6 billionth specimen of *Homo sapiens* just before the year 2000. However, in terms of population level he was remarkably correct. It was only towards the end of the twentieth century that sustainability became a real global concern. The UN World Conference on Environment and Development took place in Rio de Janeiro in 1992.

References

Anderson, P. W., K. J. Arrow, D. Pines (1988) *The Economy as an Evolving Complex System*. Redwood City, CA: Addison-Wesley Publishers.

Ansuategi, A. (1998) 'Delinking, relinking and the perception of resource scarcity', in: J. van den Bergh and M. Hofkes (eds) *Theory and Implementation of Economic Models for Sustainable Development*, pp. 165–172. Dordrecht/London: Kluwer Academic Press.

Arrow K., B. Bolin, R. Costanza, P. Dasgupta, C. Folke, C. S. Holling, B. -O. Jansson, S. Levin, K. -G. Maler, C. Perrings and D. Pimentel (1995). 'Economic growth, carrying capacity and the environment', *Science*, 268: 520–521. Reprinted in *Ecological Economics* 15 (2): 91–95.

Axelrod, R. (1984) *Evolution of Cooperation*. New York: Basic Books.

Ayres, R. U. (1994) 'Industrial metabolism: Theory and policy', in: R. U. Ayres and U. E. Simonis, *Industrial Metabolism: Restructuring for Sustainable Development*, pp. 3–21. New York: UNU Press.

Bergh, J. C. J. M. van den and J. M. Gowdy (1998) Evolutionary theories in environmental and resource economics: Approaches and applications. Unpublished paper, Dept. of Spatial Economics, Free University, Amsterdam.

Bernstein, M. A. (1993) Are developing countries 'delinking' energy demand and economic growth? EDI Working Papers, No. 93-50. Washington, DC: Economic Development Institute, World Bank.

Biervert, B. and M. Held (eds) (1992a). *Evolutorische Oekonomik: Neuerungen, Normen, Institutionen*. Frankfurt: Campus Verlag.

——— (1992b) 'Das evolutorische in der oekonomik: Eine einfuehrung', in: B. Biervert and M. Held (eds) *Evolutorische Oekonomik: Neuerungen, Normen, Institutionen*. Frankfurt: Campus Verlag.

Boserup, E. (1975) 'The impact of population growth on agricultural output', *Quarterly Journal of Economics*, pp. 257–270.

——— (1996) 'Development theory: An analytical framework and selected applications', *Population and Development Review*, 22 (3) Sept.: 505–516.

Boulding, K. E. (1970) *A Primer on Social Dynamics: History as Dialectics and Development*. London: Collier/Macmillan.

——— (1981). *Ecodynamics: A New Theory of Societal Evolution*. London: Sage.

Bruyn, S. M. de (1999). Economic growth and the environment: An empirical analysis. PhD dissertation, Free University, Amsterdam.

Bruyn, S. M. de and J. B. Opschoor (1997). 'Developments in the throughput-income relationship: Theoretical and empirical observations', *Ecological Economics*, 20 (3): 255–269.

Bruyn, S. M. de, J. C. van den Bergh and J. B. Opschoor (1998) 'Economic growth and emissions: Reconsidering the empirical basis of environmental Kuznets curves', *Ecological Economics*, 25 (2) May: 161–177.

Clark N., F. Perez-Trejo and P. Allen (1995) *Evolutionary Dynamics and Sustainable Development*. Aldershot, UK: Edward Elgar.

Chesshire, J. (1986) 'An energy-efficient future: A strategy for the UK', *Energy Policy*, 14: 395–412.

Costanza, R., L. Wainger, C. Folke and K. -G. Maeler (1993) 'Modeling complex ecological systems', in: R. Costanza, Ch. Perrings and C. J. Cleveland (eds), *The Development of Ecological Economics*. Cheltenham: Edward Elgar.

Cropper, M. and C. Griffiths (1994). 'The interaction of population growth and environmental quality', *American Economic Review*, 84: 250–254.

Daly, H. E. (1991a) 'Elements of environmental macro economics', in: R. Costanza et al. (ed.), *Ecological Economics: The science and management of sustainability*. New York: Columbia University Press.

——— (1991b). *Steady State Economics: Second Edition with New Essays*. Washington, DC: Island Press.

Darwin, Ch. (1985 [1859]) *The Origin of Species*. London: Penguin Classics.

Duncan, O. D. (1987 [1959]) 'Human ecology and population studies', in: P. M. Hauser and M. Faber, H. Niemes and G. Stephan, *Entropy, Environment and Resources: An Essay in Physico-Economics*. Berlin: Springer Verlag.

Gowdy, J. M. (1999) 'Evolution, environment and economics', in: J. C. J. M. van den Bergh (ed.), *Handbook of Environmental and Resource Economics*. Cheltenham: Edward Elgar.

Grossman, G. M. and A. B. Krueger (1996) 'The inverted U: What does it mean?' *Environment and Development Economics*, 1-1 (Feb): 119–122.

Gupta, S. and S. G. Hall (1996) 'Carbon abatement costs: An integrated approach for India', *Environment and Development Economics*, 1-1 (Feb): 41–65.

Hodgson, G. M. (1997) 'Economics and evolution and the evolution of economics', in: J. Reijnders (ed.), *Economics and Evolution*. Cheltenham: Edward Elgar.

——— (1998) 'The approach of institutional economics', *Journal of Economic Literature*, 36 (1) March: 166–192.

Jänicke M., H. Monch and M. Binder (1993) 'Ecological aspects of structural change', *Intereconomics, Review of International Trade and Development*, 28.
Klundert, Th. van der (1999) 'Economic efficiency and ethics'. *The Economist*, 147 (2) June: 127–149.
Kuznets, S. (1965) *Economic Growth and Structure: Selected Essays*. London: Heinemann Educative Books.
Leipert, Chr. (1992) 'Die normative beguensterung wirtschaftlichen wachstums durch die institutionellen bedingungen', in: B. Biervert and M. Held (eds), *Evolutorische Oekonomik: Neuerungen, Normen, Institutionen*. Frankfurt: Campus Verlag.
Magnusson, L. (ed.) (1994a) *Evolutionary and Neo-Schumpeterian Approaches to Economics*. Dordrecht: Kluwer Academic.
——— (ed.) (1994b) 'The neo-Schumpeterian and evolutionary approach to economics: An introduction', in: L. Magnusson (ed.), *Evolutionary and Neo-Schumpeterian Approaches to Economics*. Dordrecht: Kluwer Academic.
Marshall, A. (1969 [1920]) *Principles of Economics* (8th Ed.). London: Macmillan Student Editions.
Martin, K. (ed.) (1991a) *Strategies of Economic Development*. London/The Hague: Macmillan/ISS.
——— (1991b) 'Modern development theory', in: K. Martin, *Strategies of Economic Development*, pp. 27–73. London/The Hague: Macmillan/ISS.
Mill, J. S. (1988 [1848]) *Principles of Political Economy*. London: Penguin Classics.
Nelson, R. R. (1994) 'The role of firm difference in an evolutionary theory of technical advance', in: L. Magnusson (ed.), *Evolutionary and Neo-Schumpeterian Approaches to Economics*, pp. 231–242. Dordrecht: Kluwer Academic.
Nelson, R. R. and B. N. Sampat (unpublished) Making sense of institutions as a factor shaping economic performance. Columbia University, New York.
Nelson, R. R. and S. G. Winter (1982) *An Evolutionary Theory of Economic Change*. Cambridge, MA: Belknap Press of Harvard University Press.
Norgaard, R. B. (1984) 'Coevolutionary development potential', *Land Economics*, 60 (2): 160–173.
North, D. C. (1990) *Institutions, Institutional Change and Economic Performance*. Cambridge: Cambridge University Press.
——— (1991) *Structure and Change in Economic History*. London: W.W. Norton and Cie.
Odum, E. P. (1971) *Ecology*. London: Holt International.
Opschoor, J. B. (1990) 'Ecologische duurzame economische ontwikkeling: Een theoretisch idee en een weerbarstige praktijk', in: P. Nijkamp and H. Verbruggen (eds), *Het Nederlands Milieu in de Europese Ruimte*, pp. 77–126. Leiden: Stenfert Kroese, Leiden.
——— (1996) 'Sustainability, economic restructuring and social change. Inaugural Address. The Hague: Institute of Social Studies.

——— (2001) 'Economic growth, the environment and welfare: Are they compatible?' in: R. Seroa da Mota (ed.), *Environmental Economics and Policy Making in Developing Countries*, pp. 13–36. Edward Elgar: Cheltenham/Northampton.
Ormerod, P. (1994) *The Death of Economics*. London: Faber and Faber.
Ostrom, E. (1990) *Governing the Commons: The Evolution of Institutions for Collective Action*. Cambridge, MA: Cambridge University Press.
Panayotou, Th. (1993) Empirical tests and policy analysis of environmental degradation at different stages of economic development. Technology and Employment Programme, Working Paper 238. Geneva: International Labour Organization.
Persaud, B. (1997) 'The Washington Consensus revisited', in: L. Emmerij (ed.), *Economic and Social Development into the XXI Century*. Washington, DC: Inter-American Development Bank.
Ramstad, Y. (1994) 'On the nature of economic evolution', in: L. Magnusson (ed.), *Evolutionary and Neo-Schumpeterian Approaches to Economics*, pp. 65–121. Dordrecht: Kluwer Academic.
Rebergen, C., H. v.d. Vegt and J. B. Opschoor (1994) 'Economische structuur en milieudruk: Centraal Europa, 1970–1991', *Milieu*, 4: 145–153.
Reijnders, J. (ed.) (1997) *Economics and Evolution*. Cheltenham: Edward Elgar.
Rock, M. T. (1996) 'The stork, the plough, rural social structure and tropical deforestation in poor countries', *Ecological Economics*, 18(2): 113–131.
Rostow, W. W. (1956) 'The take off into self-sustained growth', *EJ*, 66: 25–48.
——— (1960) *The Stages of Economic Growth: A Non-Communist Manifesto*. Cambridge: Cambridge University Press.
Ruttan, V. W. (1998) 'Growth economics and development economics: What should development economists learn (if anything) from the new growth theory?' *Bulletin* No. 98-4. St. Paul: Economic Development Center, Department of Economics, University of Minnesota.
Selden, T. M. and D. S. Song (1994) 'Environmental quality and development: Is there a Kuznets curve for air pollution emissions?' *Journal of Environmental Economics and Management*, 27: 147–162.
Sengupta, R. (1996) *Economic Development and CO2-emission: Economy-Environment Relation and Policy Approach to Choice of Emission Standard for Climate Control.* New Delhi, India. Jawaharlal Nehru University.
Shafik, N. and S. Bandyopadhyay (1992) Economic growth and environmental quality: Time-series and cross-country evidence. World Bank Working Paper 904. Washington, DC: World Bank.
Vromen, J. J. (1997) 'Precursors, paradigmatic propositions, puzzles and prospects', in: J. Reijnders (ed.) *Economics and Evolution.* Cheltenham: Edward Elgar.
Wilkinson, R. G. (1973) *Poverty and Progress: An Ecological Model of Economic Development* [Dutch edition]. Utrecht: Aula/Het Spectrum.
WCED (World Commission on Environment and Development) (1987) *Our Common Future.* Oxford: Oxford University Press.

Watkins J. P. (1998) 'Towards a reconsideration of social evolution: Symbiosis and its implications for economics', *Journal of Economic Issues*, 32 (1) (March): 87–105.
World Bank (1992) *Development and the Environment: World Development Report 1992*. Oxford/New York: Oxford University Press.

4 Structural Dynamics and Economic Development

José Antonio Ocampo

The economic growth literature has experienced explosive growth in recent decades. Among the most important analytical innovations are the explicit recognition by the 'new' growth theories' of the role of scale economies in the growth process (as well as in international and regional analysis), the related revival of old ideas exposed by classical development economics, neo-Schumpeterian and evolutionary theories and attempts to bring institutional issues into growth analysis.[1] Extended controversies on the underlying reasons for the economic success of East Asia, as well as other successful 'late industrializers', provide parallel analytical contributions.

This richness and diversity of analytical paradigms contrasts with trends in policy design, where the triumph of a uniform liberal economics paradigm is the rule. After an era of considerable state intervention and protection of domestic markets, less interventionist, open economies were expected to provide a basis for rapid growth in the developing world. These expectations, however, have been largely frustrated so far. Latin America probably provides the most appropriate ground to test the validity of the liberal policy paradigm, its being the region of the developing world where this policy paradigm has gone the farthest in terms of implementation. The 1990s was the decade of fastest export growth in the history of the region. But at the same time it was a period of mediocre overall economic growth, indeed far below the record period of state-led industrialization in the post-war period (3.2% annually over 1990–2000 compared to 5.5% annually over 1945–80). With few exceptions, investment rates have failed to recover fully and productivity performance has been equally lagging (ECLAC 2001).

Recent contributions to economic thinking can provide useful tools to help us understand the frustrations generated by trends in policy-making and, in turn, provide the basis for alternative policies to promote economic growth. This paper attempts both tasks. It is divided into four sections. The first looks at some methodological issues and 'stylized facts'. The second focuses on the dynamics of productive structures. The third provides a very simple model of the linkages between such dynamics and overall economic and productivity growth. The fourth draws policy implications. The paper focuses on developing countries, drawing basically from the Latin American

V. FitzGerald (ed.), *Social Institutions and Economic Development,* 55–83.

experience. At an analytical level, it draws extensively from both the new and the old development literature. The elements on which the analysis is built are well known, whereas the emphasis and way these elements are put together has some novelties.

The central theme is that growth is intrinsically linked to the dynamics of productive structures and to the particular institutions that are created to support these structures. The mix of dynamic productive structures and a supportive macroeconomic environment are, in this interpretation, the clues to successful development. The broader institutional context and the adequate provision of education and infrastructure are essential background conditions, but they generally do not play a direct role in determining changes in the momentum of economic growth.

This paper draws from recent collective reflections on the development experience of Latin America by the UN Economic Commission for Latin America and the Caribbean (ECLAC 2000, 2001) and from the results of a large-scale analysis of the effects of economic reforms in the region (Stallings and Peres 2000, Katz 2000, Moguillansky and Bielschowsky 2000). It also draws from a comparative project on the economic history of Latin America in the twentieth century (Thorp 1999; Cárdenas, Ocampo and Thorp 2000a, 2000b).

Some methodological issues and stylized facts

Both time-series and cross-section analyses have identified some regularities that characterize growth processes. Productivity growth, labour skill acquisition, increased capital-labour ratios, improved infrastructure and a close association between growth and the structure of GDP and employment are among the well established regularities. There is also broad agreement on the central role of institutional development, but views vary as to the nature of this link and even on what the relevant institutions are.

The analysis of the causal links among these variables is the subject of a copious literature which involves deep methodological issues. Let me emphasize two of them. The first relates to the need to differentiate between factors that play a direct role in determining changes in growth momentum versus factors that are necessary for growth to take place but do not play a direct role in such changes; i.e. between 'proximate' and 'ultimate' causality, to use Maddison's terminology (1991: ch. 1). Institutions are probably the best case in point. Everybody would probably agree that the combination of a certain stability in the basic social contract that guarantees smooth business-labour-government relations (including the particular ideologies that serve them), a non-discretionary legal and customary system that guarantees the security of contracts, and an efficient and impartial state bureaucracy are crucial to facilitate modern, capitalist growth. Nonetheless, although in some cases they may become 'proximate' causes of growth (or of the lack of it), as in the successful reconstruction or breakdown of socio-political regimes, they generally play the role of 'background conditions' for economic growth. The specificity of growth analysis versus other branches of economic thinking relates, however, to 'proximate' causality. That is, according to the definition at the beginning of this paragraph, to those factors that play a direct role in determining changes in the growth momentum. I thus focus on these factors in this paper.

The second methodological issue relates to the fact that, as the regularities referred to above indicate, economic growth is characterized by a parallel movement in a series of economic variables: investment, savings, human capital accumulation, improved technology and systematic changes in productive structures. Yet, these variables are largely the result of economic growth rather than its determinant. Disentangling cause and effect or, in a more moderate version, leading and lagging variables, is what growth analysis is about.[2]

Empirical analysis is obviously the final test of significance of any scientific theory. The regularities mentioned above are part of the facts but, as already mentioned, they can be interpreted in different ways. I will, however, go beyond them to advance five 'stylized facts' that seem particularly important to understanding growth in the developing world. Some of these are derived from cross-section, time-series and historical analyses. Some have been seriously overlooked in contemporary growth debates.

The first is the persistence of huge inequalities in the world economy. These inequalities were determined quite early in the history of modern capitalism and have expanded through time. Empirical studies indicate that convergence in per capita incomes has been rather rare. Indeed it seems to be characteristic only of the more industrialized countries in the post-World War II period and, more particularly, in the 'golden age' period of 1950–73. It was not a characteristic of industrialized countries prior to World War II (Maddison 1991), nor has it been a characteristic of the developing world in the post-war period (Ros 2000: ch. 1). There have obviously been changes in the world hierarchy: the relative rise of the United States in the nineteenth and the first half of the twentieth century and the rise of Japan in the twentieth century, the only 'peripheral' economy that has really made it up to the top. In the developing world, there have also been some important changes: the rise of the Southern Cone Latin American countries in the late nineteenth and early twentieth century, the broader rise of Latin America in the inter-war period, and the rise of Asian newly industrialized countries in the post-war period. Yet, despite these changes in the economic landscape, the world economic hierarchy is surprisingly stable. This is reflected not only in higher incomes per capita, but also in other crucial features. Examples are the high concentration of core-technology generation in the industrialized countries and the equally high concentration there of world finance and the headquarters of multinational firms. It must be added that even the internal hierarchy in the developing world is in many ways remarkably stable.[3]

A major implication of this fact is that economic opportunities are largely determined by position within the hierarchy. Going beyond those opportunities to climb on the international ladder is a difficult task. For this reason, economic development is not a question of going through 'stages' within a uniform pattern associated with the rise in income per capita that today's industrialized countries have already gone through. Rather, it is about increasing per capita income within the restrictions imposed by the position in the world hierarchy and the internal structures in developing countries that are partly functional to it and partly the result of their historical development. This is the essential insight of the Latin American structuralist school (see e.g. Prebisch 1951, Furtado 1961) and of the literature on 'late industrialization' (Gerschenkron 1962, Amsden 2001).

A second stylized fact is that growth comes in spurts rather than as steady flows.[4] Indeed, this is the basic lesson from historical analyses. The 'inflating balloon' view of economic growth as a process by which added factors of production and steady flows of technological change smoothly increase GDP may be a useful metaphor for some purposes, but in the end it eliminates some of the most essential elements of economic development. The alternative perspective, derived from structuralist economic thinking broadly defined, views growth as a more dynamic process, in which some sectors and firms surge ahead and others fall behind, constantly changing productive structures. This process involves a repetitive phenomenon of "creative destruction" (Schumpeter 1962: ch. 8). Not all sectors have the same ability to inject dynamism into the economy, to "propagate technical progress" (Prebisch 1964). The complementarities (externalities) between enterprises and productive sectors, along with their macroeconomic and distributive effects, can produce sudden jumps in the growth process, or they can block it (Rosenstein-Rodan 1943, Taylor 1991, Ros 2000) and, in so doing, may generate successive phases of disequilibria (Hirschman 1958).

The contrast between the 'balloon' and the 'structural dynamics' views of economic growth can be understood in terms of their interpretation of one of the regularities identified in the growth literature: the tendency of per capita GDP growth to be accompanied by regular changes in the sectoral composition of output and in the patterns of international specialization (see e.g. Chenery, Robinson and Syrquin 1986; Balassa 1989). According to the balloon view, structural change may be seen as a by-product of the growth in GDP per capita and, by itself, has no explanatory power. According to the structural dynamics reading, success in structural change is the clue to economic development. The ability to constantly generate new dynamic activities, or, as I will call them, innovative activities, is the essence of the growth process. Also, structural transformations are not automatic or costless: the inability to generate new economic activities may block the development process.

The third stylized fact stresses the elasticity of factor supplies. Indeed it is difficult to understand modern economic growth in the absence of elastic factor supplies. In open economies, this is reflected in the capacity of the most successful economies to attract international capital and, when necessary, labour. The latter was more typical of the pre-World War I world. But it has also been a feature since World War II, as well as its opposite – the outward mobility of labour, particularly skilled labour, in developing economies with weak economic performance.

The internal mobility of capital and labour is even more important for developing countries. The ability to attract capital and labour is essential for dynamic, innovative activities to operate as engines of growth. More broadly, it is essential for demand factors to play a role, not only in the short-run determination of output, but also in long-run growth, an issue that lies at the root of Keynesian models of economic growth (Robinson 1962). Indeed, the central feature of pre-capitalist economies was the absence of a mobile labour force, and the ways by which dynamic activities mobilized the labour force became an essential determinant of economic and social structures (Lewis 1969; Cárdenas, Ocampo and Thorp 2000a).

Lewis (1954) provided insight into the crucial role of an elastic labour supply in development. In more contemporary versions, the interplay between labour mobility and economies of scale plays the essential role in the development process (Ros 2000, see also Krugman 1995: ch. 1). This was also a key insight of regional economic analysis. Since the origins of such analysis more than a century ago, an elastic labour supply has been seen as the basic determinant of the formation of urban and regional 'growth poles', clusters and hierarchies (for a modern version, see Fujita, Krugman and Venables 1999). That insight should obviously be extended to the analysis of international specialization and the formation of the international hierarchy we refer to above.[5] 'Vent for surplus' models of international trade, which go back to Adam Smith, provide an alternative source of elastic factor supply: the existence of underexploited natural resources (Myint 1971: ch. 5).

The fourth stylized fact stresses the dependence of long-run growth patterns on the trajectory that the economy follows, that is, path dependence (see Arthur 1994), as is evident in several development patterns. Let me mention two that are particularly relevant for the subsequent analysis. The first is the fact that, due to the role of scale economies in international trade,[6] comparative advantages are largely created, but generate specialization patterns that tend to reproduce themselves through links between productivity and prior productive experience. This may imply that in order to build up competitiveness, experience must be gained beforehand. A particular manifestation of this is the observation that successful experiences of manufacturing export growth in the developing world were generally preceded by a period of import substitution industrialization (Chenery, Robinson and Syrquin 1986). The second is the lasting effects that deep macroeconomic crises have. The debt crisis of the 1980s in Africa and Latin America is a case in point, and a similar experience may characterize some of the Asian economies that went through deep crisis in 1997–98 (Indonesia being the most remarkable example). In all these cases, the crisis clearly modified the long-term trajectory.

The fifth is closely related to the previous observation: short-term macroeconomic instability can have strong adverse effects on long-term growth. Emphasis has traditionally been on the role of inflation and balance of payments crises. Barro (1997: ch. 3) shows, in this regard, that inflation's growth-reducing effects are associated with high inflation – i.e. above 20% – with no clear patterns at lower inflation rates. The focus on inflation and balance of payments crises should not detract attention from other forms of macroeconomic instability, particularly the volatility of growth rates themselves. Indeed, to the extent that the different forms of macroeconomic instability are not strongly correlated, one form of macroeconomic stability may be chosen at the cost of instability in another sense (Ocampo 2002). The most traditional case in the developing world is that associated with price stability, when it is achieved by fixing the exchange rate, thus risking balance of payments instability. A more recent form tries to avoid some of the problems of the past. This is achieving price stability by choosing 'hard pegs', but at the risk of sharp fluctuations in real economic activity – a pattern that is strongly reminiscent of macroeconomic adjustment mechanisms during the gold standard era. Inadequate attention to a broad definition of stability and the tradeoffs involved, is certainly one of the major reasons why the return to stability, in the limited sense that this word is widely used today (as low inflation and low fiscal deficits), may not by itself generate faster economic growth.

Beyond these observations on macroeconomic stability, it is clear that policy can have strong effects on economic growth. In this regard, the strongest emphasis in the economic literature has been on the role of trade policy.[7] Nonetheless, the attempt to derive simplistic relations between open trade regimes and economic growth, and even between trade regime and export growth, runs into a dead alley (Rodríguez and Rodrik 2001).[8] The observation already mentioned, that successful experiences of manufacturing export growth in the developing world were generally preceded by periods of import substitution industrialization, indicates that simplistic generalizations are not useful. Bairoch (1993: part 1) comes to a similar conclusion regarding protection and economic growth in 'late industrializers' among the now developed countries in the pre-World War I period. Indeed he reaches the paradoxical conclusion that the fastest periods of growth of world trade prior to World War I were not those characterized by the most liberal trade regimes. Figure 1 provides additional evidence from Latin America in the 1990s. Economic growth is strongly associated with export growth, but the latter (or the former, for that matter) is not strongly associated with the trade regime.[9]

Perhaps the best conclusion that strongly emerges from comparative analyses (see e.g. Helleiner 1994) is that trade policy does matter, but that no single rule can be applied to all countries at any point in time, nor to any given country in different time periods. Protection can be a source of growth in some periods in a specific country, but can block growth in others. The same can be said of freer trade. Mixed strategies may be the optimal policy in several contexts. Openness in the world economy is obviously a decisive factor in determining what the optimal policies are at a specific point in time. This is frequently forgotten when analysing the period of state-led industrialization in the developing world; clearly import substitution made more sense in the closed world economy of the 1930s to the 1950s (and, for that matter, in the midst of the protectionist trend that characterized the industrialized world after the late nineteenth century) than in the period of gradual but incomplete opening of the industrialized world to the exports from the developing countries starting in the 1960s (Cárdenas, Ocampo and Thorp 2000b: ch. 1).

The dynamics of productive structures

The central theme of this paper is that the dynamics of productive structures, together with a supportive macroeconomic environment, are the basic direct determinants of changes in the momentum of economic growth. In terms of the previous section, the ability to constantly generate new dynamic activities is the essence of the development process. In this sense, growth is essentially the result of micro- and mesoeconomic dynamics and, as we will see, basically of the latter. Indeed, structural dynamics is basically a mesoeconomic process that affects the sectoral composition of production, intra- and inter-sectoral linkages, market structures, the functioning of factor markets and the institutions that support them. Structural and macroeconomic dynamics are indeed closely linked, as investment performance and trade balances are largely the result of structural dynamics. As I have already indicated, a facilitating broad institutional environment also serves as a 'back-

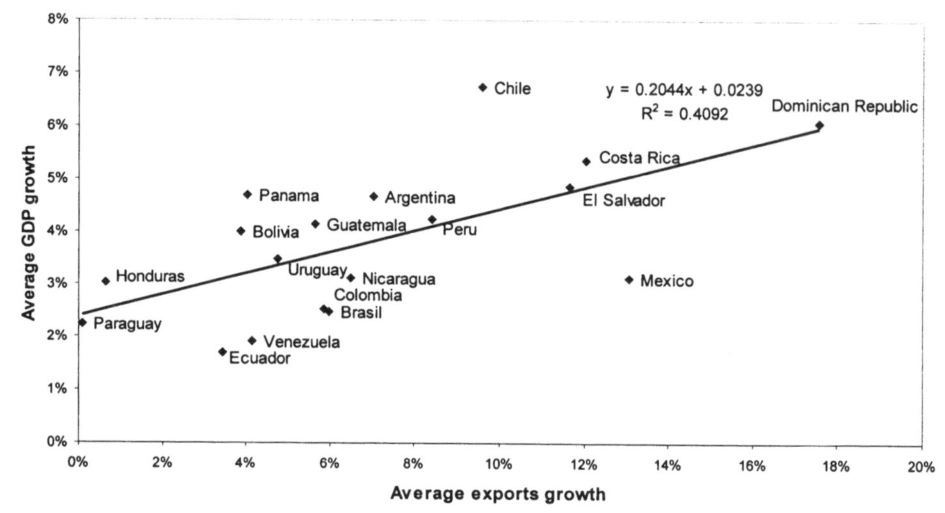

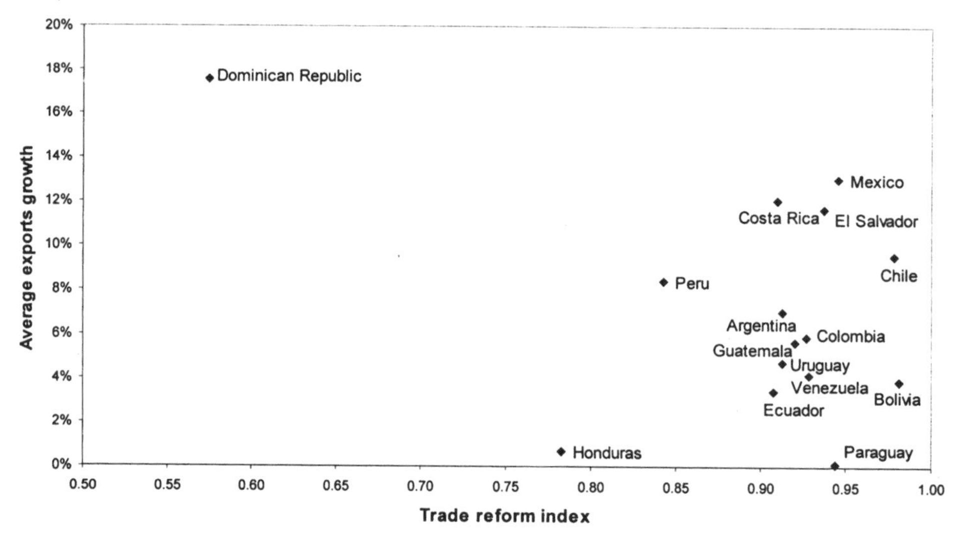

Figure 1 *Economic growth versus export growth and trade reform index in Latin America, 1990–99*

ground condition' for successful development. The supply of adequate human capital plays, as we will see, a similar role.

The dynamics of productive structures may be visualized as the interaction between four basic forces, namely (i) innovations, broadly understood as new activities and new ways of doing previously performed activities; (ii) complementarities or linkages between firms and productive sectors; (iii) the dynamic economies of scale (learning processes) that characterize both the spread of innovations and the development of complementarities; and (iv) elastic factor supplies.

The best definition of innovations, in the broad sense that I give this concept here, is that provided by Schumpeter (1961: ch. 2) almost a century ago: (i) the introduction of new goods and services or of new qualities of goods and services, (ii) the development of new productive methods or new marketing systems, (iii) the opening of new markets, (iv) the discovery of new sources of raw materials or the exploitation of previously known resources and (v) the establishment of new market structures in a given sector as the result, for example, of the creation of greater market power on the part of certain enterprises or the breaking down of dominant positions. Innovations may arise in established firms and sectors – in a constantly changing world, businesses that do not innovate tend to disappear – but many times they involve the creation of new companies and the development of new sectors of production. Thus, innovation includes both the 'creation' of enterprises, productive activities and sectors, and the 'destruction' of others that are already established.

Viewed under this broad lens, the 'innovations' of the past include the development of new export staples, the different phases of import substitution, the reorientation of import substitution sectors towards exports or the joint development of new sectors to simultaneously serve internal and external markets (a typical feature of the East Asian industrialization). In recent years, they have included the development of new activities as the result of ongoing technological revolutions (information and communications, and biotechnology), the emergence of assembly activities in developing countries (in parallel with the disintegration of such capacity in the industrialized world) and growing demand for some international services (e.g. tourism). Further examples are the increased export orientation of previously import substitution enterprises which has required special efforts to adapt product specifications and create new marketing channels, the restructuring of privatized firms and sectors, and increased access to raw materials (particularly minerals) as the result of more liberal regimes of access.

On the other hand, many 'innovations' generate 'destruction' of previous production capacities. Examples from the past include the elimination of staples as a result of changes in external conditions (new sources of raw materials or the invention of synthetic substitutes), the elimination of artisanal production unable to compete with mechanization and the reduced profitability of export activities due to the forced use of higher-cost domestic inputs or as a result of taxation and overvaluation. Examples from more recent years are the disintegration of domestic productive chains as a result of open market policies and the destruction of technological capabilities that were built up during the previous stage of development, including the dismantling of laboratories of privatized public enterprises or of private enterprises bought by multinational firms that have their own centralized research and development departments. The mix between 'creation' and 'destruction' is obviously critical. Schumpeter's (1962) term 'creative destruction' indicated that there tended to be net creation. But there may simply be 'creation' or its opposite, mere 'destruction' or a mixed case, 'destructive creation'. For growth to take place, net creative forces must prevail. So, I focus on these here.

A common feature of Schumpeter's first four forms of innovation is that they involve the creation of knowledge or, more explicitly, of the capacity to apply it to production. They thus stress the appropriation of knowledge as a source of market power. According to this view, economic development consists of the ability to create enterprises that are capable of generating and appropriating knowledge (Amsden 2001).

The incentive to innovate is provided by the extraordinary profits that can be earned by the pioneering firms that introduce technical, commercial or organizational innovations or that open new markets or find new sources of raw materials. This incentive is necessary to offset the risks and uncertainties involved in the innovator's decisions, as well as the higher costs that innovators incur due to the expense of creating and developing new know-how (including knowledge of new markets or the establishment of a trademark or goodwill in new markets), the incomplete nature of the knowledge they initially possess and the absence of the complementarities characteristic of well developed sectors. The higher costs are thus associated with the fact that two essential features that characterize the spread of innovations – complementarities and dynamic scale economies – are absent in their early stages of development.

In developing countries, innovations may be primary, but they are usually associated with the spread of new products and technologies previously created in the industrial centres. They are thus associated with the exploitation of windows of opportunity created by shifting products, technologies and organizational changes generated in the advanced countries, though these are themselves 'moving targets' (Pérez 2001). Nonetheless, as we will see below, adoption of these innovations is not a passive process: it requires active technological learning, which may itself lead to 'secondary' innovations. It also requires investments in marketing and, in many cases, the adaptation of local materials. More generally, to reduce the lag in the process of technological change that characterizes the international economic system, a more encompassing research and development strategy is needed.

Innovations are intrinsically tied to investment, as they require physical investments as well as investments in intangibles, particularly in technological learning and marketing and, to reduce the technological lag, in explicit research and development efforts. Indeed, insofar as innovative activities are the fastest growing of any economy at a specific point of time, their investment requirements are high. This fact, together with the falling investment needs that characterize established activities, implies that the overall investment rate is directly dependent on the relative weight of innovative activities and obviously on their capital intensity. High investment is thus associated with a high rate of innovation and structural change.

Complementarities are associated with the existence of networks of suppliers of goods and specialized services, marketing channels and regulatory institutions that disseminate information and provide coordination among agents. This concept summarizes the role of backward and forward linkages (Hirschman 1958) but also that of private, public or mixed institutions that are generated to reduce costs of information (e.g. on technologies and markets) and to solve the coordination failures that characterize interdependent investment decisions. Given the externalities that different economic agents generate among themselves through these mechanisms (Stewart and Ghani 1992), the existence of such complementarities determines the competitiveness – or lack of competitiveness – of productive activities. Competitiveness thus involves more than microeconomic efficiency: it is fundamentally a mesoeconomic and a system-wide feature (Fajnzylber 1990, ECLAC 1990).

In an open economy, the efficient provision of *non-tradable* inputs and specialized services, as well as of inputs and consumption goods with large transport costs (e.g. localized raw materials and perishable foodstuffs), plays a particular role in guaranteeing system-wide competitiveness. Economies of scale in the supply of some of these complementary goods and services, particularly of infrastructure, is crucial in this regard. Specialized financial services, are another vital element, particularly those, such as supply of long-term and risk capital, that are relevant to facilitate the innovative process. I return to this issue below.

As innovations mature, they give rise to significant *dynamic economies of scale*. Indeed, technical know-how must go through a learning and maturing process involving an accumulation of intangible human and organizational capital. This process is closely linked to the productive experience. Essential insights on the learning and maturing process are provided by the 'evolutionary' theories of technical change.[10] These theories emphasize the fact that technology is to a large extent tacit in nature – i.e. 'blueprints' cannot be completely spelled out. This has two major implications.

The first is that technology is incompletely available and imperfectly tradable. It is to a large extent intangible capital 'embodied' in people and organizations – i.e. human and 'social' capital. There are thus always firm-specific features that explain why companies with similar access to 'knowledge' may have quite different physical productivities, organizations and marketing strategies. This implies that, to benefit from technical knowledge, even firms that imitate or purchase it must invest in mastering the imitated or acquired technology. The efficiency with which this absorption process takes place thus determines the productivity of the firm. The implication in the case of developing countries is that, although technology is largely transferred from the industrialized countries, an active process of absorption is necessary that may involve adaptation, redesign and secondary innovations, a process that involves the active creation of human and organizational capital.

The second implication of 'tacitness' is that technology proficiency, and even technology creation, cannot be detached from productive experience – i.e. it has a strong 'learning by doing' component. Many minor and even major innovations may be the direct result of experience.[11] To such an extent, daily production and engineering have, in a sense, a 'research and development' component. This cumulative, firm-specific character of technology is the microeconomic basis of dynamic economies of scale – i.e. the link between productivity and accumulated productive experience. A particular way to express this link is that productivity increases at the firm, sectoral and national level are positively associated with economic growth (for recent evidence related to Latin America see Katz 2000). This association, which is generally known as Verdoorn or Kaldor's law, is the way we model the growth-productivity link in the following section.

A third feature of technical change, unrelated to tacitness, is that competitive pressure guarantees that innovations will be diffused or imitated. As a result, the benefits from investments in innovations are only imperfectly appropriated by the innovating firm. Technical innovations generate significant externalities and thus have mixed private-/public-good attributes. The rate of innovation thus depends on the particular balance between costs, risks, benefits and their appropriability, including legal protection through the system of intellectual property rights.

It is interesting to emphasize that similar concepts to those presented here have been developed in some versions of the new growth theory in which 'knowledge capital' is a form of 'human capital' with three specific attributes: 'embodied' in particular persons, capable of generating significant externalities and costly to acquire (Lucas 1988). However, these theories do not capture a basic corollary of firm-specificity: the fact that, at any time, sectors of production are made up of heterogeneous producers. This fact turns the concept of 'representative producer' into an irrelevant abstraction, insofar as it eliminates the very nature of competition and divergence in the growth of firms through time.

It must be emphasized that the three attributes of technical change – imperfect tradability, close association with the production experience and private-/public-good attributes – are equally characteristic of other forms of knowledge, particularly organizational and commercial know-how. The first of these, organizational know-how, probably requires no further remark. Imperfect tradability and appropiability are paramount in this case. Commercial know-how, however, plays a crucial role that tends to be left aside.[12] Yet one of the most important investments that a firm must make is in the creation of a commercial reputation and channels of commercialization and information that play a crucial role in their expansion. Moreover, familiarity with the market enables producers to modify their products and their marketing channels and helps buyers learn about suppliers, generating client relations that are important to guarantee the growth of firms.

The crucial role that these factors play is reflected in the fact that marketing departments are usually made up of high-quality personnel, indeed as qualified as those in research and development departments. The corresponding capital is essentially organizational in nature and is closely linked to commercial experience. Such experience leads to a significant reduction of transaction costs and is thus also characterized by the presence of dynamic economies of scale. Moreover, although the reputation of a particular firm can hardly be copied, its market 'knowledge' will certainly be imitated. The public-good attributes are therefore very important and play a crucial role in determining specialization. The agglomeration of producers of certain goods and services in a particular location, characteristic of both regional and international trade, is largely determined by this factor.

Institution building also shares the first two features of technological development, and by its very nature it has dominant public-good attributes. As emphasized above the two crucial services that institutions provide are the reduction of information costs and the solution to the coordination failures that characterize interdependent investment decisions. Many of the relevant institutions tend to be created directly by the private sector: producer organizations of different sorts, joint labour training facilities, specialized firms to jointly provide certain specialized services or to promote or sell the goods and services of a group of producers in certain markets and promotion agencies to encourage complementary investments. However, given their strong public-sector attributes, their services tend to be provided in suboptimal quantities. Moreover, the competitive pressure among firms is quite commonly a major obstacle to the creation and consolidation of such institutions, or a source of competing organizations of suboptimal size.

The capacity of innovations to spread depends critically on the *elasticity of factor supply*: the ability of innovative activities to attract the capital, labour and natural resources they need to expand. I already mentioned the crucial role of risk capital and the availability of finance to the innovative activities, and the fact that both have a large non-tradable component. This makes *domestic* financial development essential to guarantee a rapid rate of innovation. In the past, development, generally public-sector, banks played a crucial role in this regard in the developing world, and in many places they continue to do so. It is unclear whether privatized financial sectors will provide an adequate substitute. Private investment banking and risk capital funds are obviously the best alternatives but are highly concentrated in the developed world (and indeed, in some parts of it). In fact, some of the most important innovations in financial development in the developing world in recent decades – e.g. the pension fund revolution in Latin America – have an explicit bias against risk.

In the developing world, an elastic supply of labour is guaranteed by the structural heterogeneity that typifies its productive structure, that is the coexistence of high and low productivity activities characterized by a considerable element of underemployment or informality. Low productivity activities act as a residual sector that absorbs the excess supply of labour. The differentiation typical in dualistic models between 'traditional' and 'modern' sectors is inappropriate however. This is because the structure is certainly more heterogeneous but particularly because low productivity activities are continually being created, even to increasingly absorb skilled labour. Thus, the term 'structural heterogeneity', coined by Latin American structuralism, is clearly more appropriate. Following Ros (2000: ch. 3), three features play an essential role in the generation of an elastic supply of labour: the existence of activities with low capital requirements where the average income of the self-employed operates as a floor, the competition between these and the high productivity sectors in the provision of certain goods and services (e.g. in provision of the consumption basket of the lower-income segments of the population) and some efficiency wage features in the high productivity sectors. Depending on labour market characteristics, open unemployment and international labour migration provide additional adjustment mechanisms.

Structural heterogeneity implies that the dynamism generated by innovative activities and the strength of the linkages they generate determine the efficiency with which the aggregate labour force is used, that is, the extent of labour underemployment. At the aggregate level, this process gives rise to a Verdoorn-Kaldor productivity-growth link of similar characteristics but additional to the microeconomic links that we analysed in relation to dynamic scale economies. This link is crucial to understand the dynamics of overall productivity in the developing world. The fact that some economic agents may be approaching the technological frontier thanks to the incentives generated by a competitive environment or to their own learning effort does *not* necessarily mean that aggregate productivity will show the same degree of progress. The process itself may reduce employment, which may not be absorbed in other high productivity sectors, thus increasing under- or unemployment, adversely affecting aggregate productivity. Increased under- or unemployment may thus swamp the microeconomic gains in efficiency, generating the paradox of a growing group of 'world class' firms, with significant strides in efficiency, being accompanied by lagging rates of overall productivity growth. This was, in fact, a central feature of the Latin American panorama of the 1990s (ECLAC 2000: ch. 1, 2001).

To the extent that structural heterogeneity also involves the use of skilled labour, the more or less efficient use of the pool of educated labour may also generate an endogenous adjustment of productivity in the face of accelerating or decelerating economic growth. In this case, international labour migration provides an important additional adjustment mechanism, even more important in this case than supply of unskilled labour. This is why, although education is a crucial 'background condition' for successful economic development, it generally does not determine variations in the momentum of economic growth. A more important role may be played by the supply of the particular labour skills required by the innovative sectors, and thus the availability of corresponding public or private labour training institutions.

The concept of an elastic factor supply can be equally applied to natural resources and infrastructure. The 'vent for surplus' models mentioned earlier may be thought of as providing a similar adjustment mechanism, in which productivity is endogenously adjusted in a growing economy by exploiting previously idle or underutilized natural resources. Due to the large indivisibilities characteristic of infrastructure, particularly of transportation networks, major projects may spread their benefits over long periods. An interesting implication of this analysis is that the positive effects of infrastructure – as well as investments in education – may not only reflect the 'externalities' they generate, as emphasized in the endogenous growth literature, but also their 'fixed' or 'quasi-fixed' character, which is implicated by variable degrees of utilization, even over long periods of time.

The interplay of these four forces – innovation, complementarities, dynamic economies of scale and elastic factor supplies – is essential in the development process. We provide a simple model of this interplay in the following section. As we will see, the positive links between them reinforce each other. Innovation, if accompanied by strong complementarities and dynamic economies of scale will be reflected in the absorption of an increasing number of workers into dynamic activities. The result will be a mix of high investment, induced creation of savings and accelerated technological learning and institutional development. On the other hand, destructive forces may dominate, giving rise to adverse reinforcing effects, leading to increased structural heterogeneity as surplus labour is absorbed into less productive activities, declines in investment, the destruction of savings capacity and the loss of productive experience, all of which further accentuate the technology lag and the weakening of institutions. Intermediate cases may predominate if increased microeconomic efficiency is not accompanied by favourable mesoeconomic and macroeconomic processes.

Indeed, an interesting typology of innovations can be built on the basis of previous analysis. For example, we can distinguish between what we may call 'deep' and 'shallow' innovations. The first are those that generate strong complementarities or dynamic economies of scale, whereas the second are characterized by the weakness of these effects. In this sense, some of the innovations that the developing world has recently experienced have been shallow. A particular case is the takeover of domestic firms by multinationals if it weakens domestic demand linkages and concentrates research and development abroad. Maquila exports may have a similar character, though it can reduce underemployment and may serve as a mechanism for transmitting some organizational and marketing innovations. It could also deepen through time and gradually create domestic linkages.

The effects of deep innovations on underemployment provide another subdivision. Those characterized by strong dynamic economies of scale but weak linkages (e.g. due to high import requirements) have strong productivity growth but could maintain a large degree of dualism. Some import substitution investments of the past had that character. On the other hand, innovations with strong linkages but weak dynamic economies of scale may lead to slower direct productivity growth but to significant reduction in underemployment, which may have important indirect productivity effects. The development of labour-intensive exports is a case in point.

Path dependence is basically associated with dynamic scale economies. As it has been widely analysed in the new trade literature, such economies determine patterns of specialization that are largely self-reinforcing. Indeed, more generally, they generate processes of transformation that are context-specific, that is, they depend on unique elements that are associated with specific structural and institutional features. On the other hand, to the extent that acquired capabilities are intangible and sector-specific, rapid trade liberalization may have permanent adverse effects. Negative macroeconomic shocks could also lead to a significant loss of intangible capital. In both cases, there may be significant deadweight losses. Heymann (2000) provides an additional channel that is specifically macroeconomic: periods of rapid structural change and macroeconomic upheaval imply that past performance is no longer a useful indicator for forming expectations. Macroeconomic expectations may thus experience strong learning, a trial and error process that generates links between the short- and long-term growth paths.

A simple formalization of the growth-productivity link

The foregoing analysis can be formalized in terms of a dual link between economic growth and productivity. On the one hand, economic growth has positive effects on productivity through two Verdoorn-Kaldor mechanisms identified in the previous section: dynamic scale economies and variations in underemployment. Changes in infrastructure use also generate links of this sort. To the extent that new technology is embodied in new equipment, there is an additional mechanism: the higher rate of investment associated with faster growth also increases productivity growth. Following Kaldor (1978: chs 1–2), we call these growth-productivity links the 'technical progress function'.[13]

Figure 2 shows this relation as TT. The position of the curve depends on additional determinants of productivity growth not associated with economic growth. Some of these are derived from previous analysis: (i) the opportunity set associated with acquired technological capabilities and the position in the international hierarchy, (ii) the reaction of entrepreneurs to the alternatives ('innovativeness'), (iii) the incentives that firms face, (iv) systemic competitiveness associated with the exploitation of positive intra- and inter-sectoral externalities and (v) relevant institutions.[14]

On the other hand, technical change drives economic growth. This is the traditional link emphasized in the economic growth literature. At least three channels have been identified by different schools of economic thought. First, technical change generates new investment opportunities and, thus, drives aggregate demand. This is the Schumpeterian/Keynesian link. Second, technical change increases aggregate supply. Third, it enhances international competitiveness. This affects the trade balance and, thus, aggregate demand;

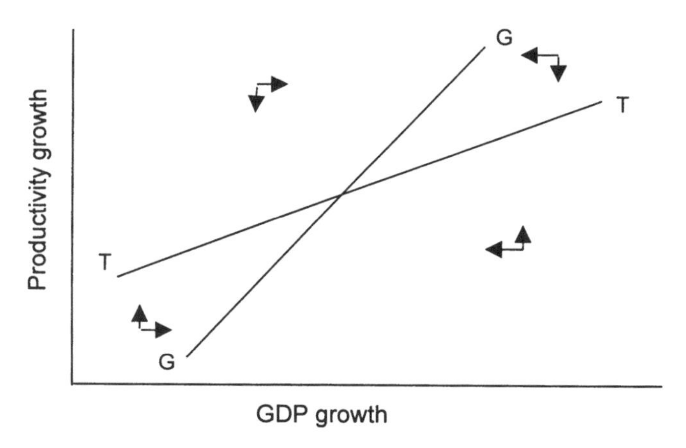

Figure 2 Productivity and GDP dynamics

and if the economy is foreign-exchange constrained, it has aggregate supply effects.

If all resources were employed in high productivity activities and full employment was guaranteed, growth would be supply-determined; the relevant determinants would then be factor growth and technical change. However, as indicated in the previous sections, these conditions are not necessarily guaranteed. This gives room for aggregate demand effects to play a determining role. As indicated above, these are associated with investments that respond to opportunities generated by innovations and to improvements in the trade balance generated by increased competitiveness. It is important to note that, whereas the productivity (supply) effects of complementarities affect the TT function, the corresponding demand effects are captured in this Keynesian relation. Finally, if the economy is foreign-exchange constrained, GDP growth would be determined by export growth and changes in import dependence. Figure 2 shows these productivity-growth relations as GG.

As both curves have positive slopes, the effects that they summarize reinforce each other, generating alternatively win-win but also possible lose-lose situations. A stable equilibrium exists when TT is flatter than GG, as Figure 2 shows. Given the determinants of the technical progress function, this is likely if the following conditions prevail: (i) the most profitable innovations are exploited first, (ii) dynamic economies of scale and embodied technical progress are not too strong, (iii) productivity heterogeneity is moderate and (iv) fixed factors are not very important in the long run. Its slope is then likely to be significantly less than one. However, under significant initial structural disequilibria, factors (iii) and (iv) could generate high slopes. On the other hand, in Keynesian and foreign-exchange gap models, the slope of GG depends on the elasticity of investment, exports and imports to productivity; if they are relatively inelastic, the corresponding schedules will be steep. In any case, as is typical in models with increasing economies of scale, multiple and unstable equilibria may exist. Under unstable equilibria, any displacement from an initial saddle point leads to either explosive 'virtuous' or 'vicious' growth paths. Also, nothing guarantees that equilibrium is always reached at positive rates of growth.

It is important to emphasize that the relations shown are taken to be long-term in character.[15] However, as many of the processes we are analysing are time-bound, the possible steady-state properties of the model are really uninteresting. Indeed, innovations may be seen as 'spurts' that tend to weaken through time. So, a new wave of innovation shifts the TT function up and turns it steeper, accelerating both productivity and income growth. However, as a particular wave of innovation becomes fully exploited and its structural effects fully transmitted, TT shifts down and becomes flatter. Productivity and GDP growth slow too. If the GG function also shifts down (weakened 'animal spirits'), the slowdown would become even sharper.

Macroeconomic instability affects growth through the GG relation, adversely affecting investment and exports (if export incentives are unstable). Following the considerations introduced in the first section of this paper, *any* form of instability matters. The GG function shifts leftward. The micro/meso links summarized in the technical progress function now multiply the macro effects of this shift. Macroeconomic instability becomes a very adverse environment for these micro/mesoeconomic processes. Positive movements associated with improvements in investment finance have the opposite effect, even aside from their effects in terms of facilitating innovations, to which we referred in the previous section.

This simple framework may be used to analyse the effects of trade liberalization. For that purpose, we have to settle on, first of all, what is the specific relation that exists between competition and innovation. In this regard, we must recall that, contrary to the Schumpeterian tradition, which emphasizes that large firms' ability to internalize benefits from innovation may generate a positive link between market concentration and productivity, the neoclassical tradition sees the lack of competitive pressure as adverse to productivity. This view highlights the fact that managers of large firms may be inclined to appropriate part of the monopoly power they hold in the form of 'leisure' (reduced efforts to minimize costs). The corresponding deviation from cost minimization ('x-inefficiency') must be weighted against the traditional Schumpeterian benefits of size. We should emphasized, however, that this neoclassical assumption implies that firms are *not* profit-maximizers (in traditional terms, that entrepreneurs are 'satisficing' rather than optimizing agents), a presumption that has major theoretical implications in other areas which are not usually recognized.[16]

If the neoclassical assumption were correct, opening the economy to competition would potentially displace the TT function up. Liberalization unleashes in this case a degree of 'innovativeness' that the more protected, state-interventionist environments of the past repressed. However, there is more that matters. The destruction of domestic linkages and previous technological capabilities has the opposite effect. Specialization in activities with weaker dynamic economies of scale tends to make the TT function flatter. If firms shrink, their capacity to incur fixed costs associated with innovative activities also declines. One way to express this point of view is that, although the microeconomic effects of competition on productivity growth may be positive, the mesoeconomic (structural) factors may be adverse. The net effect is thus unclear.

Figure 3 provides three possible outcomes. In Case A, the neoclassical effects on TT are strong and prevail over the weaker adverse movements of the

Case A: Strong TT, weak GG effects

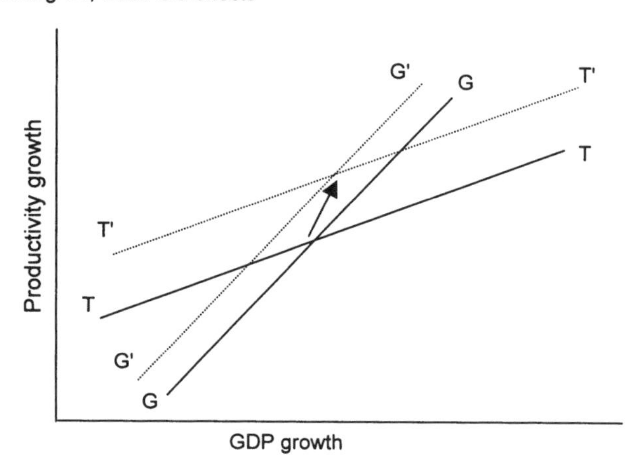

Case B: Weak favorable TT, strong GG effects

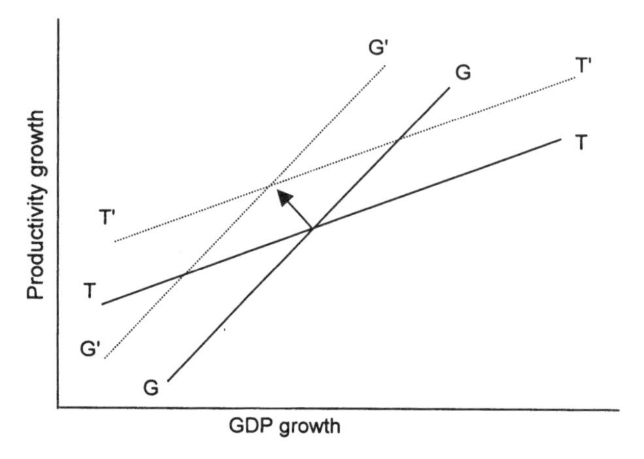

Case C: Adverse TT and GG effects

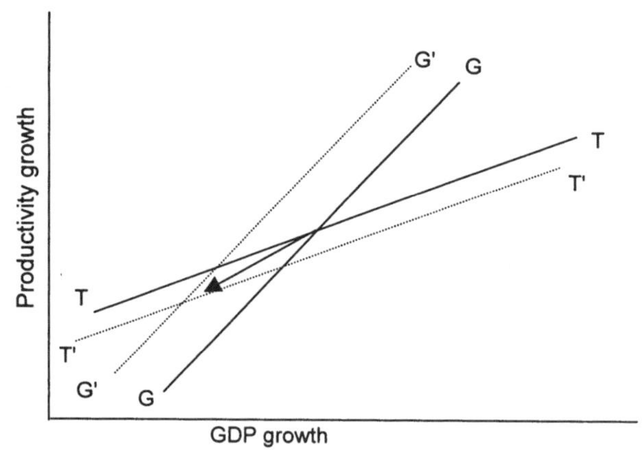

Figure 3 Relation between TT and GG, three scenarios

GG function. Both GDP and productivity growth speed up. In Case B, neoclassical effects on TT continue to prevail, but GG effects are strong. Productivity growth speeds up but overall economic growth slows. An implication of this is that under- and unemployment increase sharply. In Case C, adverse structural effects on TT prevail over the positive effects of competition, generating a reduction in both GDP and productivity growth. It must be emphasized, again, that this is not inconsistent with positive microeconomic effects of competition on productivity. If those effects are strong, under- and unemployment will increase sharply.

In either a Keynesian or a foreign-exchange constrained economy, the GG function clearly moves to the left. This is reflected in the Latin American case by the fact that the same rate of growth is now associated with a larger trade deficit – or, alternatively, the same trade deficit is associated with a lower rate of growth (Figure 4a). On the other hand, the low ratio of GDP to export growth (Figure 4b) indicates that export activities are much less integrated domestically. According to previous remarks, this may be a factor adversely affecting the TT function.

The data presented in Figure 5 indicate that, for most Latin American countries, labour productivity growth was indeed weaker in the 1990s than in the period of state-led industrialization (1945–80). The exceptions are the three countries of the Southern Cone and the Dominican Republic. Of these, only Chile and the Dominican Republic have strong performance. The Argentinean productivity figures are affected by the strong positive recovery experienced in 1991–94, after hyperinflation was stopped. Since 1994, however, its productivity and GDP performance have been relatively weak. In practice, Case C has thus been the most common result. Since some of the microeconomic effects were indeed strong, as reflected in the rapid growth of productivity of some firms in the face of increased competition, the adjustment path also involved increased under- and unemployment, a common feature in the Latin American panorama of the 1990s (ECLAC 2000, 2001).

Policy implications

The previous analysis indicates that institutions that guarantee stability in the basic social contract, the protection of contracts and an efficient state bureaucracy are certainly important to economic growth. But they play the role of 'background conditions' that, by themselves, are unlikely to affect growth momentum. A similar argument can be applied to the formation of human capital and the development of infrastructure.

The clue to rapid growth in the developing world is thus the combination of *strategies aimed at the transformation of productive structures* with

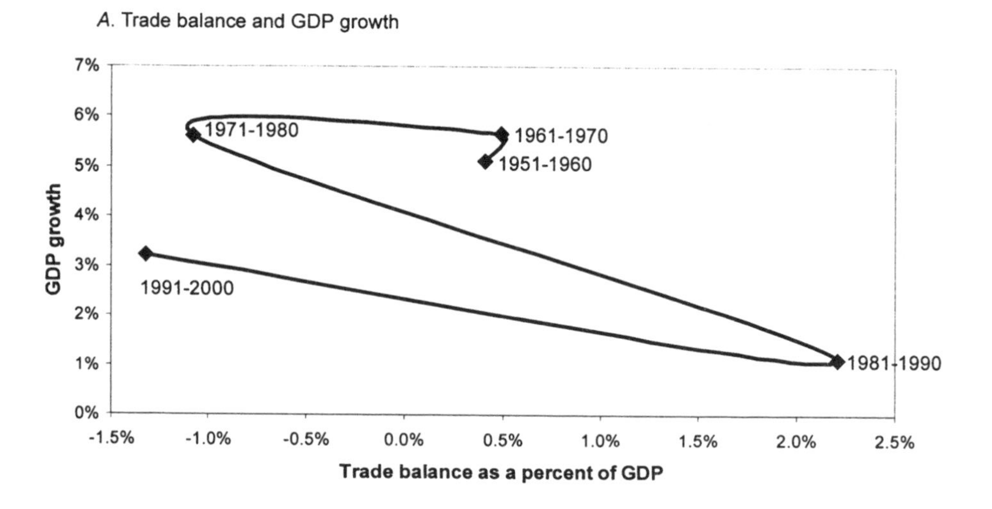

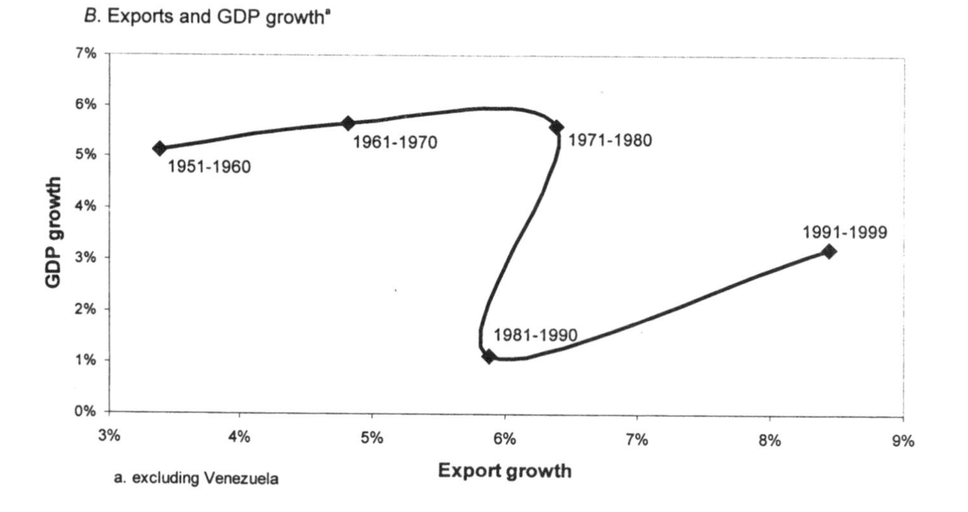

Figure 4 Trade and economic growth in Latin America, 1950–2000

Source: ECLAC.

appropriate *macroeconomic and financial environments* that guarantee, in particular, macroeconomic stability in the broad sense of the term (in overall price levels, in real economic activity and in crucial relative prices). Since, according to the views presented here, innovations and investment are tied, this view coincides with Rodrik's (1999) call for a "domestic capital accumulation strategy" to kick-start growth, mixed with appropriate macroeconomic management. It is important to emphasize that this is the combination that

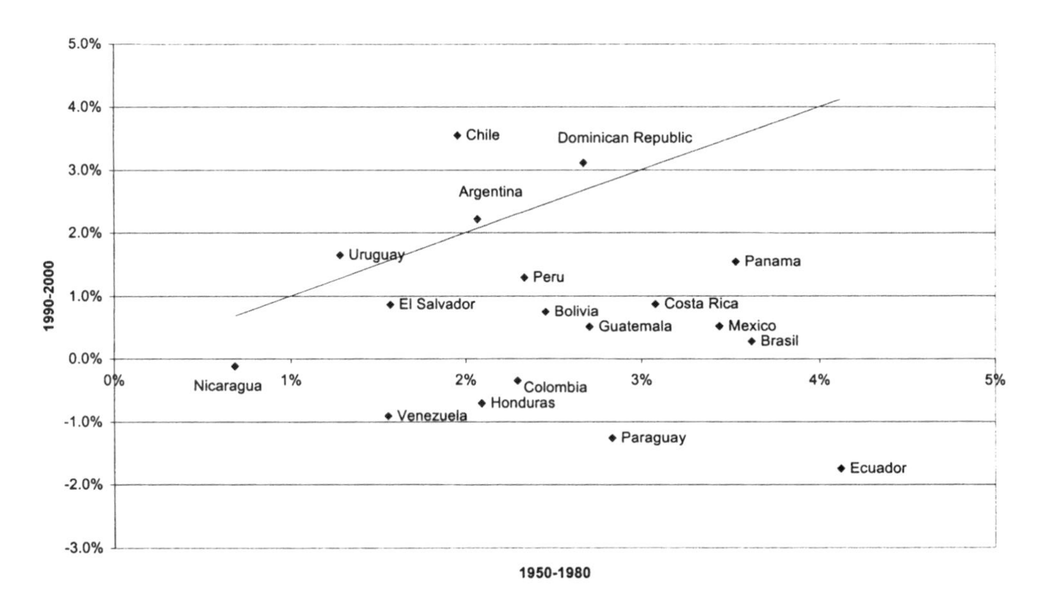

Figure 5 Labour productivity growth, 1950–80 versus 1990–2000

explains the rapid growth of the Asian economies. The vigorous growth that took place in Latin America during the period of state-led industrialization was also the product of a structural change strategy that was based initially on import substitution, but increasingly on 'mixed' models combining import substitution with export promotion (Cárdenas, Ocampo and Thorp 2000b: ch. 1). Unlike the Asian countries, the lack of adequate macroeconomic conditions bred the debt crisis of the 1980s that led to a sharp break in the growth pattern.

The focus on the structural dynamics helps us to clearly identify the specific policy areas that authorities should target to accelerate economic growth: encourage innovation (in the broad sense of the term), complementarities and dynamic learning processes. Under current international conditions, we can identify four essential characteristics of productive transformation strategies that should serve as the framework for these policies. In the first place, the emphasis should be on integrating the developing countries into the world economy, and thus on developing regional integration processes and integrated export sectors. In the second place, there must be a proper balance between individual initiative, which is decisive for getting a dynamic process of innovation started, and the establishment of public incentives and adequate coordination systems. Thirdly, all incentives should be granted on the basis of performance, in order to generate 'reciprocal control mechanisms', to borrow Amsden's (2001) terminology. Fourthly, a broad mix of public and private institutions should be considered, with each country developing the combination that best suits its own particular needs. Finally, these policies should be applied in a macroeconomic and financial context that is conducive to the restructuring of existing capacity and that encourages productive investment.

A complex issue relates to the framework of international rules, especially those of the World Trade Organization. In this regard, although priority should be given to taking advantage of the manoeuvring room provided under existing agreements, there is a strong sense that more opportunities should be available to the authorities of developing countries, who were too narrowly restricted after the Uruguay Round. In particular, they should be allowed to apply selective policies and performance criteria to encourage innovation and create the complementarities that are essential to development.

The facts prove wrong the assumption that dynamic productive structures, and the institutions that support them, are automatic results of market mechanisms. In Latin America, the weakening of the public and private institutions that had been established to support productive development during the period of state-led industrialization was a central feature of the 'lost decade' of the 1980s. In the 1990s these institutions were weakened further as a result of explicit policy decisions. Some institutions have emerged around free trade zones, production clusters and the promotion of small and medium-size enterprises, but generally, again, as a result of explicit policy. The suboptimal development of institutions to support productive development has thus become the most important direct institutional deficiency affecting economic growth. This institutional deficiency is probably not very important if growth is to remain at current levels. However, institutional development is crucial if Latin America aims for the rapid pace structural change – including the penetration into dynamic technology-intensive sectors – that will be needed for it to gradually bridge the gap with the industrialized world.

Notes

1. The recent literature is extensive. Among the most useful contributions are Romer (1986), Lucas (1988), Taylor (1991), Barro and Sala-i-Martin (1995), Nelson (1996), Aghion and Howitt (1998) and Ros (2000).
2. There may also be intermediate alternatives. Some factors may not 'cause' growth in the sense of accelerating growth momentum, but can block it.

3. Some important differences in income per capita within Latin America were established in the early twentieth century and have been stable since then (Cárdenas, Ocampo and Thorp 2000a: ch. 1).
4. A contrast that has many elements in common with this is the contrast between 'yeast' versus 'mushrooms' as the essential feature of the growth process (Harberger 1998).
5. This was recognized in the seminal work by Ohlin (1933), but only another part of his thinking, that on the effects of relative factor supplies on comparative advantage, made it into mainstream analysis.
6. See in particular Krugman (1990), Grossman and Helpman (1991) and, in relation to developing countries, Ocampo (1986).
7. See the survey on the literature of the 1980s by Edwards (1993) and the critical survey on that of the 1990s by Rodríguez and Rodrik (2000).
8. My own contributions to this debate are in UNCTAD (1992: part 3, ch. 1).
9. The corresponding index takes into account two factors: average tariffs and their dispersion. The broader issue of the association of economic growth and reforms in Latin America are explored in detail in Stallings and Peres (2000). Escaith and Morley (2000) provide econometric evidence. A major conclusion of these studies is that the elements of the liberalization packages had opposite effects on economic growth, largely neutralizing themselves, and the short- and long-term effects often had opposite signs.
10. See in particular Nelson (1996) and Dosi et al. (1988) and, with respect to developing countries, Katz (1987), Lall (1990) and Katz and Kosakoff (2000).
11. This may also apply to technology creation. In this sense, the probability of major innovations, even when they are the result of explicit research and development efforts, also depends on the accumulated technological knowledge and productive experience of firms.
12. For a treatment of international marketing issues see Keesing and Lall (1992).
13. Productivity increases are understood here as the 'residual' in traditional calculations, including scale economies (both static and dynamic, as it is in practice impossible to differentiate the effects of one or the other over time). Also, as already indicated, they refer to aggregate movements, which are affected not only by microeconomic efficiency but also by variations in underemployment and the relative use of fixed factors.
14. For a further analysis of some of these issues see Lall (1990).
15. There are short-run relations of the sort shown. However, in the short run, productivity is strongly dependent on capacity utilization. The TT function is, thus, not independent of GG, if the latter is affected by aggregate demand or by import availability.

16. See Rodrik (1992).

References

Aghion, Philippe and Peter Howitt (1998) *Endogenous Growth Theory*. Cambridge, MA: MIT Press.

Amsden, Alice (2001) *The Rise of the Rest: Non-Western Economies' Ascent in World Markets*. Oxford: Oxford University Press.

Arthur, W. Brian (1994) *Increasing Returns and Path Dependence in the Economy*. Ann Arbor: University of Michigan Press.

Bairoch, Paul (1993) *Economics and World History: Myths and Paradoxes*. Chicago: University of Chicago Press.

Balassa, Bela (1989) *Comparative Advantage Trade Policy and Economic Development*. New York: New York University Press.

Barro, Robert J. (1997) *Determinants of Economic Growth: A Cross-Country Empirical Study*. Cambridge, MA: MIT Press.

Barro, Robert J. and Xavier Sala-i-Martin (1995) *Economic Growth*. Columbus: McGraw-Hill.

Boyer, Robert (1990) *The Regulation School: A Critical Introduction*. New York: Columbia University Press.

Cárdenas, Enrique, José Antonio Ocampo and Rosemary Thorp (eds) (2000a) 'The export age: The Latin American economies in the late nineteenth and early twentieth centuries', in: *An Economic History of Twentieth Century Latin America*, Vol. 1. Basingstoke: Palgrave Press and St. Martins.

——— (2000b) *Industrialisation and the State in Latin America: The Post War Years; An Economic History of Twentieth Century Latin America*, Vol. 3. Basingstoke: Palgrave Press and St. Martins.

ECLAC (2001) *Luces y Sombras: América Latina y el Caribe en los Años Noventa*, Bogotá: CEPAL/Alfaomega.

——— (2000) *Equity, Development and Citizenship*. Santiago: United Nations.

——— (1990) *Changing Production Patterns with Social Equity: The Prime Task of Latin American and Caribbean Development in the 1990s*. March. Santiago: United Nations.

Chenery, Hollis, Sherman Robinson and Moshe Syrquin (1986) *Industrialization and Growth: A Comparative Study*. New York: The World Bank/Oxford University Press.

Dosi, Giovanni, Christopher Freeman, Richard Nelson, Gerald Silverberg and Luc Soete (eds) (1988) *Technical Change and Economic Theory*. London and New York: Pinter Publishers.

Edwards, Sebastián (1993) 'Openness, trade liberalization, and growth in developing countries', *Journal of Economic Literature*, 31 (3) September: 1358–1393.

Escaith, Hubert and Samuel Morley (2000) 'The impact of structural reforms on growth in Latin America and the Caribbean: An empirical estimation', Macroeconomía del Desarrollo Series No. 1, November. Santiago: ECLAC.
Fajnzylber, Fernando (1990) 'Industrialization in Latin America: From the "black box" to the "empty box"', Cuadernos de la CEPAL No. 60. Santiago: CEPAL.
Fujita, Masahisa, Paul Krugman and Anthony J. Venables (1999) *The Spatial Economy: Cities, Regions and International Trade*, Cambridge, MA: MIT Press.
Furtado, Celso (1961) *Desarrollo y Subdesarrollo.* Buenos Aires: Editorial Universitaria de Buenos Aires.
Gerschenkron, A. (1962) *Economic Backwardness in Historical Perspective.* Cambridge, MA: Harvard University Press.
Grossman, Gene M. and Elhanan Helpman (1991) *Innovation and Growth in the Global Economy.* Cambridge, MA: MIT Press.
Harberger, Arnold C. (1998) 'A vision of the growth process', *The American Economic Review*, 88 (1) March: 1–32.
Helleiner, Gerald K. (ed.) (1994) *Trade Policy and Industrialization in Turbulent Times.* New York: Routledge.
——— (ed.) (1992) *Trade Policy, Industrialization, and Development: New Perspectives.* New York: Oxford University Press.
Heymann, Daniel (2000) Major macroeconomic upsets, expectations and policy responses. CEPAL Review No. 70 (LC/G.2095-P). Santiago: CEPAL.
Hirschman, Albert O. (1958) *The Strategy of Economic Development.* New Haven: Yale University Press.
Kaldor, Nicholas (1978) *Further Essays on Economic Theory.* London: Duckworth.
Katz, Jorge (2000) *Reformas Estructurales, Productividad y Conducta Tecnológica.* Santiago: Economic Commission for Latin America and the Caribbean (ECLAC)/Fondo de Cultura Económica.
——— (1987) 'Domestic technology generation in LDCs: A review of research findings', in: Jorge Katz (ed.), *Technology Generation in Latin American Manufacturing Industries*. London: Macmillan.
Katz, Jorge and Bernardo Kosakoff (2000) 'Technological learning, institutional building, and the microeconomics of import substitution', in: Enrique Cárdenas, José Antonio Ocampo and Rosemary Thorp (eds) *Industrialisation and the State in Latin America: The Post War Years; An Economic History of Twentieth Century Latin America, Vol. 3*. Basingstoke: Palgrave Press and St. Martins.
Keesing, Donald B. and Sanjaya Lall (1992) 'Marketing manufactured exports from developing countries: Learning sequences and public support', in: Gerald K. Helleiner (ed.) *Trade Policy, Industrialization, and Development: New Perspectives*. New York: Oxford University Press.
Krugman, Paul (1995) Development, Geography and Economic Trade. Cambridge, Mass: MIT Press.
——— (1990) Rethinking International Trade. Cambridge, Mass: MIT Press.
Lall, Sanjaya (1990) *Building Industrial Competitiveness in Developing Countries*. Paris: OECD Development Center.

Lewis, W. Arthur (1969) Aspects of Tropical Trade, 1883–1965. Stockholm: Almqvist & Wicksell, Wicksell Lectures.

——— (1954) 'Economic development with unlimited supplies of labor', *Manchester School of Economic and Social Studies*, 22 (May).

Lucas, Robert E., Jr. (1988) 'On the mechanics of economic development', *Journal of Monetary Economics*, 22 (1) July: 3–42.

Maddison, Angus (1991) *Dynamic Forces in Capitalist Development: A Long-Run Comparative View.* Oxford: Oxford University Press.

Myint, H. (1971) *Economic Theory and the Underdeveloped Countries.* New York: Oxford University Press.

Moguillansky, Graciela (1999) *La Inversión en Chile: ¿El Fin de un Ciclo en Expansión?* Santiago: Fondo de Cultura Económica/Economic Commission for Latin America and the Caribbean (ECLAC).

Moguillansky, Graciela and Ricardo Bielshowsky (2000) *La Inversión en un Proceso de Cambio Estructural: América Latina en los Noventa.* Santiago: Economic Commission for Latin America and the Caribbean (ECLAC)/Fondo de Cultura Económica.

Nelson, Richard (1996) *The Sources of Economic Growth.* Cambridge, MA: Harvard University Press.

Ocampo, José Antonio (2002) 'Developing countries' anti-cyclical policies in a globalized world', in: Amitava Dutt and Jamie Ros (eds) Development Economics and Structuralist Macroeconomics: Essays in Honour of Lance Taylor, Aldershot, UK: Edward Elgar.

——— (1986) 'New developments in trade theory and LDCs', *Journal of Development Economics*, 22 (1) June: 129–170.

Ohlin, B. (1933) *Interregional and International Trade.* Cambridge, MA: Harvard University Press.

Pérez, Carlota (2001) 'Technological change and opportunities for development as a moving target', *CEPAL Review* No. 75, Santiago.

Prebisch, Raúl (1964) *Nueva Política Comercial para el Desarrollo.* Mexico City: Fondo de Cultura Económica.

——— (1951) Theoretical and practical problems of economic growth. (E/CN.12/221) UN publication sales No. 52.II.G.1. Mexico City: Economic Commission for Latin America.

Robinson, Joan (1962) *Essays in the Theory of Economic Growth.* London: Macmillan.

Rodríguez, Francisco and Dani Rodrik (2001) 'Trade policy and economic growth: a skeptic's guide to the cross-national evidence', in: Ben S. Bernanke and Kenneth Rogoff (eds.) *NBER Macroeconomics Annual 2000*, Vol. 15. Cambridge, MA: MIT Press.

Rodrik, Dani (1999) *The New Global Economy and Developing Countries: Making Openness Work.* Washington, DC: Overseas Development Council.

——— (1992) 'Closing the productivity gap: Does trade liberalization really help?' in: Gerald K. Helleiner (ed.), *Trade Policy, Industrialization, and Development: New Perspectives.* New York: Oxford University Press.

Romer, P. M. (1986) 'Increasing returns and long-run growth', *Journal of Political Economy*, 94 (5): 1002–1037.
Ros, Jaime (2000) *Development Theory and the Economics of Growth*. Ann Arbor: University of Michigan Press.
Rosenstein-Rodan, P. N. (1943) 'Problems of industrialization of Eastern and South-Eastern Europe', *The Economic Journal*, 53 (June–September): 202.
Schumpeter, Joseph (1962) *Capitalism, Socialism and Democracy*, Third Edition. New York: Harper Torchbooks.
——— (1961) *The Theory of Economic Development*. Oxford: Oxford University Press.
Stallings, Barbara and Wilson Peres (2000) *Growth, Employment and Equity: the Impact of the Economic Reforms in Latin America and the Caribbean*. Santiago: Economic Commission for Latin America and the Caribbean (ECLAC)/Fondo de Cultura Económica.
Stewart, Frances and Ejaz Ghani (1992) 'Externalities, development, and trade', in: Gerald K. Helleiner (ed.), *Trade Policy, Industrialization, and Development: New Perspectives*. New York: Oxford University Press.
Taylor, Lance (1991) *Income Distribution, Inflation, and Growth. Lectures on Structuralist Macroeconomic Theory*. Cambridge, MA: MIT Press.
Thorp, Rosemary (1999) *Progress, Poverty and Exclusion: An Economic History of Latin America in the 20th Century*. Baltimore: Johns Hopkins University Press.
UNCTAD (1992) *Trade and Development Report*. Geneva: UN Conference on Trade and Development.

5 Economic Reforms, Development and Distribution: Were the Founding Fathers of Development Theory Right?

Rob Vos

Introduction

"Development and, more specifically, an industrial revolution may be a rather unpleasant process." Kurt Martin wrote this about a decade ago in his marvellous essay on modern development theory (Martin 1991: 47). While development should eventually lead to poverty reduction, the welfare of important parts of society may lag behind during the period of transitional growth to an industrial society. Historical evidence shows that development has been an inegalitarian process in many ways. Early development theory is often seen to have provided the theoretical foundations for what some – in line with Kuznets – consider an empirical regularity. Lewis' model for one predicts growing inequality during the period of transition growth as an increasing profit share in national income is required to finance industrial investment (Lewis 1954).

All this could lead us to believe that the founding fathers of classical development theory were firm believers in 'trickle down' mechanisms. Of course, much development policy debate since the 1970s has centred around the question of how the development process could be made less unpleasant and, indeed, how growth and poverty reduction could go hand in hand. The fact of the matter is that the founding fathers too were very much concerned with income distribution and, in fact, with the creation of conditions such that growth and redistribution would go together. Lewis (1955) and Prebisch (1961), for instance, showed this awareness quite forcefully in the 1950s and early 1960s. Their argument focused on the primary distribution of income and the pattern of growth rather than on redistribution through taxation and social spending, but most theorists probably favoured a combination of the two.

Development economics as it was laid out by its founding fathers, such as Kurt Martin, Rosenstein-Rodan, Nurkse, Prebisch, Leibenstein and others, very much emphasized the obstacles to industrialization and physical and human capital formation in developing countries, with income inequality

V. FitzGerald (ed.), Social Institutions and Economic Development, 85–99.

and poverty being both cause and effect. The obstacles that had to be overcome to achieve development included foreign-exchange bottlenecks, lack of the social overhead capital (including human capital) needed for the generation of positive vertical externalities and increasing returns to scale, and insufficient domestic linkages to stimulate employment and income generation supportive of modern growth.

Modern growth economics and, more in particular, the new, endogenous growth theories try to address similar issues. But curiously enough, they ignore by and large the contributions of development theory. This may be typical of many economists who do not bother much with the history of their science. As Jean Baptiste Say said long ago, "The more perfect the science, the shorter its history" (quoted in Blaug 2001: 146). So if economics is real science, economists should not concern themselves with its founders and pretend they are creating new ideas and make *tabula rasa* of earlier contributions. I will not argue that the early thinkers gave us all the right answers to yesterday's and today's questions. Yet, I will emphasize that the issues they addressed are still relevant today and therefore unduly forgotten.

The links between income distribution, market size and industrialization have been central to development economics from its foundations. Productivity growth in agriculture might lead to an increase in the size of the market for manufactured goods, making it profitable for manufacturing firms to shift to an increasing returns to scale technology. In this story, income distribution becomes pivotal for economic growth. Too much equality might lead to insufficient savings and investment finance, whereas too much inequality would lead to a lack of wage-goods demand. Both ills would lead to development traps of too little capital formation and/or smaller demand for manufactured goods, leading in turn to a delay in industrialization. The notions of 'too much inequality' and 'too much equality' are of course somewhat fuzzy, and they would suggest that one could define an optimal income distribution or an optimal degree of inequality consistent with a maximum sustainable growth rate. To my knowledge the concept of an 'optimal income distribution' has never been developed, although from an analytical point of view development theory might have benefited from it.

The development debate took a different route of course, one reason being that the critics of classical development economics saw that economic openness would present the solution to the problems of industrialization. Critics, such as Bhagwati (1985: 299), emphasized that the alleged 'export-elasticity pessimism' provided the weak link in the argument, which led the pioneers of development economics to focus on closed economies and misconceived policy advice of protectionism and import substitution.[1] If the problem arises from a lack of demand, opening the economy to international trade would be the way out of the vicious circle. However, the point of founding fathers such as Rosenstein-Rodan was not merely the problem of low elasticities, trade pessimism and lack of demand. The more essential point was the need to create technological externalities (such as 'learning by doing' and ensuring adequate social overhead capital), which would leave the overall argument intact also in the open economy.[2] In today's language, the heart of the matter then becomes whether the economy possesses adequate infrastructure, human capital and entrepreneurial skill (learning by doing) to take advantage of the opportunities provided by the global economy. Having put it this way, I am inclined to ask, "Now what's new endogenous growth theory?"

The policy environment has changed dramatically of course. The founding fathers saw economic planning, aid and trade protection as important instruments to overcome the perceived development bottlenecks. Today's conventional wisdom is that globalization and free movement of commodity and capital flows have become the prime movers of growth and development. Building on recent research conducted in a large number of Latin American countries I argue that economic liberalization has not brought the type of 'Big Push' that its advocates had hoped for. Further, the 'post-modern' growth process seems to have exacerbated inequality rather than having resolved it.

The move towards liberalization in Latin America

Looking back from the end of the twentieth century, the most striking aspect of economic policy in developing economies during the last 10–15 years has been the spread of packages aimed at liberalizing the balance of payments, on both the current and capital account. Dramatic leaps toward external openness were made throughout Latin America, Eastern Europe, Asia and parts of Africa. Together with large but highly volatile foreign capital movements (often but not always in connection with privatization of state-owned enterprises), this wave of trade and financial deregulation redefined the external environment for a major part of the non-industrialized world. In Latin America, the stabilization and structural adjustment efforts that immediately followed the debt crisis of the early 1980s focused mainly on fiscal and monetary adjustment and realignment of exchange rates. Then, in the late 1980s and early 1990s, came drastic reductions in trade restrictions and domestic and external financial liberalization, almost simultaneously in most countries. Steps were also taken toward restructuring tax systems and deregulating labour markets.

All these changes are very recent. It will take time before their full effects on growth, employment, income distribution and poverty can be fully assessed. Still, external liberalization marks a dramatic switch in development policies away from the traditional regime of widespread state controls and import substituting industrialization, much of which – it can be held – found justification in the insights of the pioneers of development. One would expect to see major consequences. The old regime to a large extent was built on the infant-industry argument to create 'learning-by-doing' externalities and enhance Hirschman-type domestic linkages so as to lay the foundations of a sustainable growth process.

Import substitution did yield moderate to high growth for a prolonged period of time, as GDP growth averaged over 6% per annum and productivity (measured as output per worker) doubled between 1950 and 1970 (Stallings and Peres 2000). Despite the relatively successful growth performance, the pioneers of development economics were among the first to observe the flaws of the policy regime even before the economies ran out of steam and macroeconomic problems started to mount.[3] Broadly, the protectionist regime was criticized for failing to promote efficient and competitive industrial production (and thereby providing a source of 'structuralist inflation'), for creating insufficient employment and for failing to reduce income inequality. Sectoral balance and income distribution formed a central element in the critique: the protectionist policies had biased relative prices in favour of capital-intensive industrial production causing employment creation to lag behind population growth and skewing income distribution against wage earners and farmers. Widening inequalities set a limit to the growth of the domestic market and thus to further growth. The solutions had to be found in redistribution policies as much as economic opening. As said, in the final event, full economic liberalization became the dominant paradigm of the new policy regime, characterizing the end of classical development economics as a factor of influence in shaping development policies.

A fundamental question now is whether the liberalization of trade and capital flows will be better at meeting the developmental goals of growth, equity and poverty reduction. Will a world system in which national economies are highly integrated in commodity and capital markets (in terms of increased transaction flows and tendencies toward price equalization) promote equality and reduce poverty?

The reforms have been justified by expected increases in efficiency and output growth. Yet governments and international institutions promoting them have been less explicit about the distributional consequences. A predominant view is that liberalization is likely to lead to better economic performance, at least in the medium to long run. Even if there are adverse transitional impacts, they can be cushioned by social policies, and in any case after some time they will be outweighed by more rapid growth.

This policy view basically stems from supply-side arguments. The purpose of trade reform is to switch production away from non-tradables and inefficient import-substitutes toward exportables in which countries have a comparative advantage. Presumed full employment of all resources – labour included – enables such a switch to be made painlessly. Opening of the capital account is supposed to bring financial inflows that will stimulate investment and productivity growth. A recent defence based on cross-country regressions for Latin America (Londoño and Székely 1998) argues that equity is positively related to growth and investment. In turn, these are asserted to be positively related to structural reforms, so liberalization is seen to support low-income groups.

This story contrasts with findings of many other studies which, referring in particular to the effects of trade reforms, find that opening domestic markets to external competition is associated with greater wage inequality (Robbins 1996, Wood 1994, 1997, Ocampo and Taylor 1998). Berry (1998) and Bulmer-Thomas (1996) corroborate this proposition with data for a range of Latin American countries, observing a shift in technology in favour of more capital- and skill-intensive production consistent with a rise in wage differentials. However, the evidence stems essentially from the 1980s and possibly captures more of the effects of short-run adjustment policies than of trade and capital account liberalization.

While there may be important supply-side effects to trade reforms, one should not overlook the effects of aggregate demand on growth and distribution, which is a central theme of development economics, and the impact of capital inflows on relative prices, which is an issue underestimated by the pioneers. The import substitution model relied on the expansion of internal markets with rising real wages as part of the strategy. Under the new regime, controlling wage costs has moved to centre stage. As long as there is enough productivity growth and no substantial displacement of workers, wage restraint need not be a problem because output expansion could create space for growth of real incomes. But if wage levels are seriously reduced or workers with high consumption propensities lose their jobs, contraction of domestic demand could cut labour income in sectors that produce for the domestic market. Income inequality could rise if displaced unskilled workers end up in informal services for which there is a declining demand.

Rising capital inflows following liberalization tend to lead to real exchange rate appreciation, which could offset liberalization's incentives for traded goods production and force greater reductions in real wage costs. On the demand side, capital inflows may stimulate aggregate spending through increased domestic investment (either directly or through credit expansion) and lower savings (credit expansion triggering a consumption boom). However, aggregate demand expansion may prove to be short-lived if the consequent widening of the external balance is unsustainable and volatility of short-term capital inflows and lack of regulatory control puts the domestic financial system at risk.

The thrust of these observations is that the effects of balance of payments liberalization on growth, employment and income distribution come from a complex set of interactions involving both the supply and the demand sides of the economy. Income redistribution and major shifts in relative prices are endogenous to the process, and there can be no facile conclusions about the effects of liberalization.

Growth, distribution and poverty in Latin America: Recurring problems

While there are no easy conclusions, evidence from a comparative study of the post-liberalization performance of 17 Latin American and Caribbean economies during the 1990s suggests that the diverging outcomes are closely associated with the precise issues that concerned the pioneers of development.[4] The differences in outcomes have much to do with the links between market size, employment and income distribution and the adequacy of social overhead capital (including human capital investment).

While most Latin American countries achieved moderate growth rates in the 1990s, with few exceptions it is hard to speak of a strong and sustained recovery from the dismal performance of the 1980s. What's more, toward the end of the decade growth tapered off in many countries due to emerging domestic financial crises – as was the case in Paraguay, Colombia and Ecuador – or external events. Adverse foreign shocks included the impact of the Asian crisis on capital flows to Brazil with spill-over effects on neighbouring countries, particularly Argentina, and the effects of falling export earnings for most primary exporting economies due to plummeting commodity prices. While also for Latin America it is true that poverty falls with growth (see Figure 1), there are important deviations from the trend line strongly associated with specific macroeconomic conditions and, more in particular, with the pattern of growth.

Macroeconomic conditions

Let us first look at some of these macroeconomic conditions. Particularly in the first half of the 1990s, capital inflows to most countries increased substantially and brought both aggregate demand growth and real exchange rate appreciation (with a few exceptions, see below). The latter outcome has been consistent with reductions in inflation, which helped support higher average real wages in most countries. The surge in capital inflows produced expansionary macroeconomic cycles and the associated real wage increases lifted

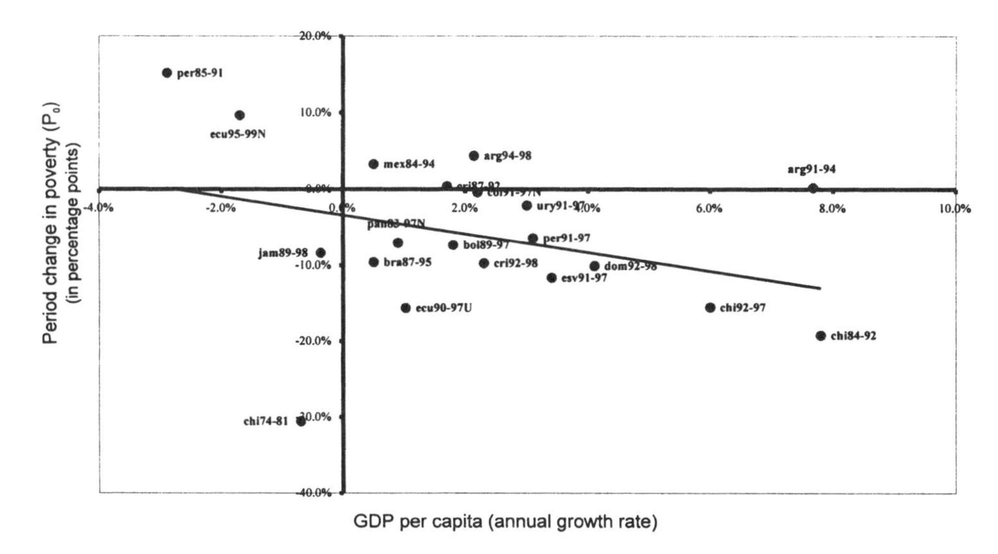

Figure 1 Growth and poverty in Latin America in the 1990s

domestic market constraints. Growth would accelerate and poverty would fall during such episodes, but rather than constituting a 'Big Push', the overall picture is one of macroeconomic 'go-stop' cycles (Taylor and Vos 2001), as wages and aggregate demand strongly contract as capital inflows slow. This is further corroborated by the fact that exports in only a few country cases provided the engine of growth during the 1990s. Private spending in most cases proved to be the major source of growth, with consumption growth more often than investment being the major driving force (Table 1).

Remarkably, export-led growth has been closely associated with macroeconomic policy regimes which maintained either relatively competitive exchange rates or a credible system of export incentives, or both. These cases by and large coincide with those that also managed relatively strong poverty reduction effects (such as Chile, El Salvador and Guatemala) as export growth in these cases induced strong employment growth. This finding may require further in-depth analysis, but at first sight should raise some scepticism about the virtues of fixed exchange rate regimes or dollarization, which are popular policy options these days in the region. Also, to the extent that export promotion schemes facilitate externalities to export producers (either by providing social overhead infrastructure or reducing costs), one might see shades of the pioneers of development theories. Further, capital flow volatility has been damaging, generating high economic and social costs (De Feranti et al. 2000, World Bank 2001).

Hence, while we label it differently these days, regulation of financial markets and related institutional reforms may well be seen as an essential in-

Table 1 Factors of Growth in Latin America in the 1980s (by aggregate demand decomposition)

	Country	Periods	Principal source of demand growth	Aggregate demand growth (% per year)
1	Argentina	1990–94	Private consumption boom	8.9
		1995–96	Private demand contraction	–4.6
		1996–98	Private demand (C,I) recovery	6.5
2	Bolivia	1980–85	Private consumption and gov. spending	–1.5
		1986–89	Export led	2.1
		1990–97	Export led	4.8
3	Brazil	1982–86	Government spending and exports	–0.9
		1987–91	Government spending	3.0
		1992–94	Private spending and gov. spending	0.9
		1994–97	Private investment and consumption	5.2
4	Chile	1970–74	Private and government consumption	1.0
		1976–81	Consumption squeeze, export growth	9.4
		1985–89	Investment, exports	8.4
		1990–97	Investment, exports	9.4
5	Colombia	1990–92	Exports and government spending	2.2
		1992–95	Private consumption boom	9.6
		1995–98	Private demand contraction	1.5
6	Costa Rica	1985–91	Export led	5.7
		1992–98	Export led	6.5
7	Cuba	1989–93	Private demand squeeze	–13.7
		1994–98	Public spending and export recovery	7.0
8	Dominican Rep.	1993–99	Private demand and export led	7.5
9	Ecuador	1988–91	Private demand	4.4
		1992–98	Export led	2.9
10	El Salvador	1990–95	Investment and export	8.2
		1996–97	Export	0.1
11	Guatemala	1986–91	Consumption led	3.4
		1991–98	Consumption led	5.0
12	Jamaica	1980–89	Private consumption led	2.0
		1990–92	Export led	8.1
		1993–98	Private demand and export contraction	–3.1
13	Mexico	1988–94	Consumption boom	5.5
		1994–95	Crisis and cons. squeeze	–7.8
		1996–98	Investment recovery	8.3
14	Panama	1986–90	Crisis: private demand contraction	–5.4
		1990–94	Private demand and exports	5.7
		1994–98	Exports and private demand	4.9
15	Paraguay	1988–91	Private demand expansion	6.7
		1992–94	Private demand expansion	10.8
		1995–98	Private demand and export contraction	–0.6
16	Peru	1986–90	Collapse private demand	–1.9
		1991–97	Private demand recovery	5.6
17	Uruguay	1986–90	Export led, private demand squeeze	2.9
		1990–94	Private demand expansion	8.4
		1994–97	Private demand and exports	4.4

Source: Taylor and Vos (2001).

gredient of what Rosenstein-Rodan and others called the social overhead capital required to generate the positive externalities that would enable achievement of the 'Big Push'.

Patterns of growth and inequality

Turning to the pattern of growth and income distribution, the most generalizable result is that the inequality of primary incomes increased almost across the board during the 1990s (Table 2). When separated from other influences, it turns out that trade liberalization has been the major cause of this rise in inequality (see Ganuza, Barros and Vos 2001). Trade liberalization came with a skill-twist. Looking deeper inside sectoral adjustment patterns, one finds that the drive towards efficiency gains has led to the adoption of more skill-intensive technologies in many instances driving abundant unskilled workers into unemployment or low-paid informal sector employment. Shortage of human capital, which one may define as another 'Big Push' element, has driven up income inequality. Virtually without exception, wage differentials between skilled and unskilled workers rose in Latin America during the post-liberalization period. Excess labour was typically absorbed in the non-traded, informal trade and services sectors (as in Bolivia, Colombia, Costa Rica, Ecuador, Panama and Peru), or – as happened in a few cases – traditional agriculture served as a sponge for the labour market (Panama in the late 1980s, Guatemala and Mexico).

The sectoral patterns have not been uniform. In Argentina, for instance, productivity increases in the traded goods sector affected workers of all skill levels. Wage rigidity being greater for unskilled workers, there was a reduction in earnings inequality in the sector. But greater inequality in Argentina was due to rising income concentration in the non-traded sector along with the greater skill-intensity of new investment and to the rise of unemployment in the traded goods sector. By contrast, in Mexico reorganization of manufacturing production was found to be a major source of greater skill demand, pushing up wage inequality in the traded goods sector with many of the displaced workers absorbed by agriculture, at least until 1994. In Brazil, productivity growth produced employment losses in the manufacturing sector. Labour demand fell for everyone in modern manufacturing, but skilled workers suffered the most. Real hourly wages fell for both skilled and unskilled workers in modern industry, but slightly less for unskilled workers, showing – as in Argentina – greater rigidity in wage adjustment at the lower end; hence skilled-unskilled income differentials showed a slight decline. As indicated, in most other cases such productivity growth in the traded goods sector pushed up skill differentials in that sector along with the gap between formal and informal sector workers.

Table 2 Growth and Inequality in Latin America in the 1980s

Change After Liberalization		INEQUALITY		
		Overall Primary Incomes		
		Rising Inequality	**Decreasing Inequality**	**Unchanged**
GROWTH	High (>5%)	ARG (91–94, 96–98) CHI (76–81, 84–92) COL (91–95) DR (91–98) PERU(91–97)	CHI (92–97) ESV(91–97) PAN(90–94)	URY (90–97)
	Moderate (2–5%)	BOL (89–97) BRA(87–94) CRI (92–98) ECU (90–97) MEX (88–94) PAN(94–98) PRY (88–91, 92–94)	BRA(94–97) CRI (87–92) CUB (94–98)	URY (86–90)
	Low (0–2%)	COL (95–98) ECU(95–99) MEX (85–87) PRY (95–98)	JAM(89–98)	
	Negative (< 0%)	CUB (89–93) MEX (94–95)		

Change After Liberalization		INEQUALITY		
		Skill Differentials		
		Rising Inequality	**Decreasing Inequality**	**Unchanged**
GROWTH	High (>5%)	ARG (91–94, 96–98) CHI (76–81, 84–92) COL (92–95) DR (91–98) ESV (90–97) PAN(90–94) PERU (91–98) URY (90–97)	CHI (92–97)	
	Moderate (2–5%)	BOL (89–97) BRA (92–94) CRI (85–91, 92–98) ECU (90–97) MEX (88–94) PRY (88–91, 92–94)	BRA (94–97) URY (86–90)	
	Low (0–2%)	COL (95–98) JAM (90–92) MEX (85–87) PAN(94–98) PRY (95–98)		
	Negative (< 0%)	JAM(93–98) MEX (94–95)		BRA (87–91)

Source: Taylor and Vos (2001).

The picture is not entirely gloomy as far as primary income distribution is concerned. In El Salvador, rapid growth of employment for unskilled workers, particularly in export sectors, offset widening between-group skill differentials. In Chile, overall labour market tightening was probably the main factor behind the reduction of wage differentials in the 1990s. In Brazil, elimination of hyperinflation and labour demand shifts toward the unskilled were factors underlying the dampening of primary income differentials. Trends have also been influenced by minimum wage policies, as in Ecuador, where upward adjustments in the minimum wage allowed for a temporary decline in earnings inequality (1992–95) despite an overall rising trend (1990–98). In Jamaica, real exchange rate appreciation implied a relative price shift in favour of non-traded activities, which in an overall stagnant economy attracted many unskilled workers from rural areas and the agricultural sector. As urban living standards are generally higher and real wages were allowed to grow, the sectoral employment shift explains the reduction in overall income inequality among workers despite the widening wage gap between the skilled and unskilled.

As Figure 2 shows, rising per worker differentials do not necessarily translate into rising inequality and poverty at a household level. The cases of rising inequality clearly predominate once more (east of the vertical axis), but so do episodes where poverty fell during the 1990s (south of the horizontal axis). Economic growth evidently helped reduce poverty, also where liberalization pushed toward greater inequality. In only a few cases – particularly Chile, Guatemala and El Salvador – was poverty reduction associated with moderate to strong export-led growth and falling inequality. In most other cases, growth recovery following a surge in capital inflows allowed for an expansion of aggregate demand and sufficient overall employment growth or a rise in real wages to produce a reduction in poverty. In Mexico and Argen-

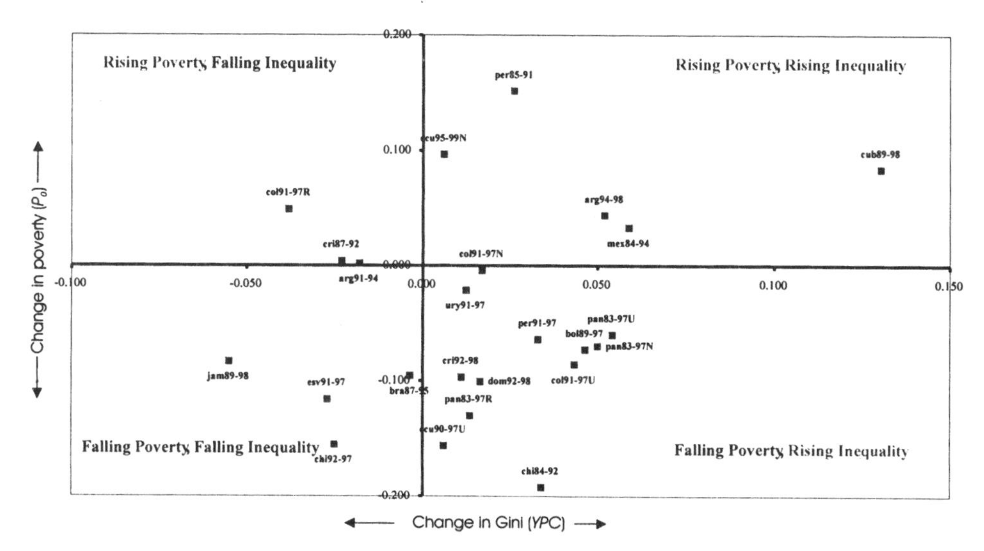

Figure 2 Poverty and inequality (of per capita household income) in Latin America before and after liberalization

tina, the rise in inequality, particularly associated with labour demand shifts favouring skilled workers and employment shifts into informal activities or unemployment, caused a rise in poverty despite positive per capita growth. In other cases, changing labour market conditions triggered strong labour supply responses, including rising female participation, as in for example, Panama and urban Ecuador. Elsewhere, emigrant remittances (Central America, Dominican Republic, Cuba) or social security transfers (e.g. Costa Rica) had a strong positive influence on reduction of poverty and inequality at the household level.

Conclusions

The development context has changed dramatically. Yet when it comes to questions of development and distribution, the founding fathers emphasized the right issues. In line with the spirit of the time they were probably affected by too much trade pessimism and perhaps by a too great belief in the virtues of planning and protectionism. It is fair to say though that none of the pioneers believed in autarkic development, and all saw that eventually the full benefits of trade could be reaped by opening up the economy, not blindly, but in a fashion that would lead developing economies out of development traps. Development still could be rather 'unpleasant' as Kurt Martin said. The pioneers had no clear recipe for making development a more pleasant process during the transition. Rather, they warned for too much inequality as much as for too much equity, both of which could hamper modern economic growth. Here is the problem that economists have been unable to resolve to date, despite more sophisticated analytical methods and much improved data by which to study income distribution and poverty.

The challenge is to make sure we bring back into the equation the fundamental concepts of sectoral balance, social overhead capital (broadly defined) and vertical technological linkages. This does not imply that we should give a clarion call for a return to wide-ranging trade intervention policies and import substitution. The insights from old development theory have already taught us the high costs associated with the existence of fragmented markets, inadequate institutional frameworks to guide market processes and large income inequality. Our research and policy advice should focus on such issues.

Notes

1. The critique has been restated more recently by other less orthodox economists like Krugman (1992) and Stiglitz (1992).
2. Murphy, Shleifer and Vishny (1989a, b) and Ros (2000) provide modern formalized reinterpretations of the basic notions put forward by Rosenstein-Rodan. They argue that the critique of Bhagwati and others in fact is valid only in the case of 'horizontal pecuniary externalities', that is, demand spillovers across final producers of traded goods. In a closed economy, the profitability of a shoe factory, to take Bhagwati's example, depends not only on its own production function, but also on what other producers do, such as textile producers. When the economy opens up, the adoption of new, modern techniques in textile production will still affect the domestic demand for shoes, but the domestic demand for shoes would no longer form a constraint on the profitability of modern techniques in shoe production. However, when vertical externalities pose the central problem then Rosenstein-Rodan's fear of a development trap remains. A shoe factory producing in isolation faces high-cost intermediate inputs such as services and infrastructure, yielding multiple disequilibria of the sort Rosenstein-Rodan hinted at. Murphy et al. and Ros modelled this.
3. See for instance Prebisch (1961, 1963) and Hirschman (1968).
4. The findings of the study will soon be published in Spanish (Ganuza et al. 2001) and English (Vos, Taylor and Paes de Barros 2001).

References

Berry, Albert (ed.) (1998) *Poverty, Economic Reform, and Income Distribution in Latin America.* London: Lynne Rienner.

Bhagwati, Jagdish (1985) *Essays in Development Economics*, Vol. 1. Cambridge, MA: MIT Press.

Blaug, Mark (2001) 'No history of ideas please, we're economists', *Journal of Economic Perspectives,* 15: 145–164.

Bulmer-Thomas, Victor (ed.) (1996) *The New Economic Model in Latin America and its Impact on Income Distribution and Poverty.* London: Macmillan.

De Feranti, David, Guillermo Perry, Indermit Gill and Luis Servén (2000) 'Securing our future in the global economy'. World Bank Latin American and Caribbean Studies. Washington, DC: World Bank.

Ganuza, Enrique, Lance Taylor, Ricardo Paes de Barros and Rob Vos (eds) (2001) *Liberalización de la Balanza de Pagos en América Latina y el Caribe: Efectos sobre el Empleo, la Distribución y la Pobreza.* Buenos Aires: Ediciones Universidad de Buenos Aires.

Ganuza, Enrique, Ricardo Paes de Barros and Rob Vos (2001) 'Effects of liberalization on poverty and inequality', in: Rob Vos, Lance Taylor and Ricardo Paes de

Barros (eds), *Economic Liberalization and Income Distribution: Latin America in the 1990s.* Cheltenham: Edward Elgar.

Hirschman, Albert O. (1968) 'The political economy of import-substituting industrialization in Latin America', *Quarterly Journal of Economics,* 82: 2–32.

Krugman, Paul (1992) 'Toward a counter-counterrevolution in development theory', in: *Proceedings of the World Bank Annual Conference on Development Economics, 1992*, pp. 15–38. Washington, DC: World Bank.

Lewis, W. Arthur (1954) 'Economic development with unlimited supplies of labour', *Manchester School of Economic and Social Studies,* 26: 1–31.

——— (1955) *The Theory of Economic Growth.* London: Allen and Unwin.

——— (1976) 'Development and distribution', in: Alec Cairncross and Mohinder Puri (eds), *Employment, Income Distribution, and Development Strategy: Essays in Honour of Hans Singer*, pp. 26–42. London: Macmillan.

Londoño, Juan Luis and Miguel Székely (1998) 'Sorpresas distributivas después de una década de reformas', *Pensamiento Iberoamericano-Revista de Económica Política* (Special Issue).

Martin, Kurt (1991) 'Modern development theory', in: Kurt Martin (ed.), *Strategies of Economic Development. Readings in the Political Economy of Industrialization*, pp. 27–73. London: Macmillan.

Meier, Gerald M. and Dudley Seers (eds) (1984) *Pioneers in Development.* New York, Oxford: Oxford University Press (for the World Bank).

Murphy, Kevin, Andrei Shleifer and Rob Vishny (1989a) 'Industrialization and the big push', *Journal of Political Economy*, 97: 1003–1026.

——— (1989a) 'Income distribution, market size, and industrialization', *Quarterly Journal of Economics*, 104: 537–564.

Ocampo, José Antonio and Lance Taylor (1998) 'Trade liberalisation in developing economies: Modest benefits but problems with productivity growth, macro prices and income distribution', *Economic Journal*, 108: 1523–1546.

Prebisch, Raúl (1961) 'Economic development or monetary stability: A false dilemma', *Economic Bulletin of Latin America*, 6 (1): 1–25.

——— (1963) *Towards a Dynamic Development Policy for Latin America.* New York: United Nations.

——— (1984) 'Five stages in my thinking on development', in: Gerald M. Meier and Dudley Seers (eds), *Pioneers in Development*, pp. 175–91. New York, Oxford: Oxford University Press (for the World Bank).

Robbins, Donald (1996) *HOS Hits Facts: Facts Win: Evidence on Trade and Wages in the Developing World.* Cambridge, MA: Harvard Institute for International Development.

Rosenstein-Rodan, Paul (1943) 'Problems of industrialization in eastern and south-eastern Europe', *Economic Journal*, 53: 202–211.

——— (1984) 'Natura facit saltum: Analysis of the disequilibrium growth process', in: Gerald M. Meier and Dudley Seers (eds), *Pioneers in Development*, pp. 207–221. New York, Oxford: Oxford University Press (for the World Bank).

Ros, Jaime (2000) *Development Theory and the Economics of Growth*. Ann Arbor: University of Michigan Press.

Stallings, Barbara and Wilson Peres (2000) *Growth, Employment, and Equity: The Impact of Economic Reforms in Latin America and the Caribbean*, Washington, DC: The Brookings Institute and ECLAC.

Stiglitz, Joseph (1992) 'Comment on "Toward a counter-counterrevolution in development theory" by Krugman', in: *Proceedings of the World Bank Annual Conference on Development Economics, 1992*, pp. 39–49. Washington, DC: World Bank.

Taylor, Lance and Rob Vos (2001) 'Balance of payments liberalization in Latin America: Effects on growth, distribution and poverty', in: Rob Vos, Lance Taylor and Ricardo Paes de Barros (eds), *Economic Liberalization and Income Distribution: Latin America in the 1990s*. Cheltenham: Edward Elgar.

Vos, Rob, Lance Taylor and Ricardo Paes de Barros (eds) (2001) *Economic Liberalization and Income Distribution: Latin America in the 1990s*. Cheltenham: Edward Elgar.

Wood, Adrian (1994) *North-South Trade, Employment and Inequality: Changing Fortunes in a Skill-Driven World*. Oxford: Clarendon Press.

——— (1997) 'Openness and wage inequality in developing countries: The Latin American challenge to East Asian conventional wisdom', *World Bank Economic Review*, 11: 33–57.

World Bank (2001) *Global Development Finance 2001*. Washington, DC: World Bank.

6 Why Groups Matter[1]

Frances Stewart

Introduction

This paper explores the role of groups, an important issue which tends to be treated inadequately by those working within the neoclassical paradigm. By groups, I mean collections of people who act and take decisions together rather than individually. Groups, therefore, cover a wide range of organizations and activities: they extend to firms in the private sector, to central and local government, to workers, consumers and producers, cooperatives or collectives and even to families.[2]

Darwin concluded that participation in groups is essential for the survival of the species. Whatever we want to achieve – building pyramids, university education, playing football, producing cars, travelling in planes, practising religion or politics – we have to act in groups or we fail. Babies would never survive if they were not born into families. As individuals we lack power and judgement; as groups we can literally move mountains.

Groups obviously play an important part in our social life[3] – one indicator of this being the considerable evidence that isolated individuals suffer psychological traumas. Yet, groups are equally important in economic life. Adam Smith focused on the process of production in his famous example of the gains from specialization. Dividing up production and specializing, Smith found that 10 people produced over 4,000 times more pins together than if they worked separately. Less noticed, but nonetheless essential, was the fact that the 10 were employed together in 'a small manufactory', that is, as a group. Working separately as individuals and trading each part with one another would have involved intolerable transactions costs and thus ruled out specialization. In the modern economy, intra-group activities greatly exceed inter-group, with intra-firm transactions of multinational corporations accounting for as much as a quarter of manufacturing trade (UN 1992). The public sector accounts for a third or more of national income and within-family activities account for perhaps the equivalent of over half of measured national income.[4] Strictly market transactions account for only a fraction of total economic activity, even in the more market-oriented economies. In such successful market economies as Japan and Korea, companies

V. FitzGerald (ed.), *Social Institutions and Economic Development,* 101–124.

grouped into Zaibatsu (pre-World War II Japan) and Keiretsu (post-World War II Japan) or Chaebol (Korea) account for the dominant portion of industrial production. Elsewhere, such as Taiwan, family groups play an equivalent role.

Despite the central importance of groups in economic life, only individuals, not groups, have a role in the Walrasian system. Even the (mythical) auctioneer is an individual, not a collective entity. In this paradigm, some groups are regarded as welfare distorting, acting to restrain trade as Adam Smith argued, or to seek rents and perform 'directly unproductive activities', according to Krueger and Bhagwati, while others, such as families and firms, are mostly viewed as quasi-individuals. Of course, recent developments – the 'new institutional economics' which includes the transactions cost approach to the firm, theories of collective action and public choice theory – do pay attention to groups. The first two attribute a positive functionalist role for them, the last a mainly negative one. But for neoclassical economists (though not, of course, for sociologists)[5] groups remain for the most part an anomaly: the rational maximizing individual continues to be the central actor, and group activity remains something to be explained as a response to 'market failure' not as a natural phenomenon. Public-sector groups are viewed with suspicion, as being a reflection of competing interest groups, not as servants of the public interest.

Communitarians, in contrast, recognize the central role of groups in human life, arguing that "a basic observation of sociology and psychology is that the individual and the community 'penetrate' one another and require one another, and that individuals are not able to function without deep links to others" (Ezioni 1993: 65). Their view of human motivation differs fundamentally from the neoclassical approach, representing motivation as "I and we" as Etzioni puts it. That is, individuals are assumed to identify with others, and consequently not only to maximize their own satisfaction as in the neoclassical model.[6] In contrast to the neoclassical view, groups are believed to emerge naturally in human society, to play a crucial role in economic life, including in the market, and do not need to be 'explained' as a response to market failures.

Communitarian thinkers, perhaps, have a rather romantic notion of both human motivation and the role of groups. Analysis must recognize that groups enhance our power for ill, as well as good. At one extreme, groups are responsible for developments in Bosnia and for the chaos in Somalia. In less extreme contexts, but still with negative effects, groups of businesspeople can elevate prices and prevent innovation; groups of workers may be responsible for restrictions on activities and entry so that productivity is depressed and opportunities are unequally distributed.

This paper provides a sketchy overview of the role of groups with economic functions in development, drawing on both neoclassical and communitarian views. The purpose is to explore different types of group behaviour and to identify reasons why some groups appear to work well and others badly, always bearing in mind a central assumption: that humans invariably operate in groups so group behaviour is not an exceptional but a central phenomenon which we must explore if we are to understand development.

The paper is organized as follows: the next section puts forward some simple definitions and categories to help in classifying and understanding group behaviour. Case studies are then presented of different types of group behaviour, contrasting successes and failures in several areas. The final section draws some conclusions. The main focus, particularly in the empirical section, is on public-sector and community groups, not firms or families. This is not because firms and families are regarded as unimportant, but to limit the canvas, which is already overambitious.

The nature and functioning of groups

What are groups?

Groups are relationships among individuals working for common purposes. The unit typically takes group rather than individual decisions vis-à-vis those outside the group.[7] Most individuals participate in a number of different groups, such as their family, work organization and social organizations, which can give rise to conflicting loyalties, but may also generate complementarities.

Groups may be classified as formal or informal. Formal groups are clearly delineated, subject to agreed membership rules and rules of operation. Often they are termed 'organizations'. The firm, a trade association, a trade union, or a government organization are examples. They often have agreed and written rules, some legally enforceable. In contrast, membership of informal groups may be unclear, sporadic and varying. The family, the neighbourhood and some interest groups are examples. Informal groups are also subject to rules, often implicit.

Both formal and informal groups are subject to values, norms, conventions and traditions. These are themselves informal, emerge historically and change over time. For shorthand, we refer to them as norms. They condition how a group operates. In the business literature, a firm's norms are known as its culture.[8] Some norms are specific to the group in question, whereas others emerge from and are shared by the society in which the group is located.

Rules and norms apply, of course, not only to relations within groups but also to interactions between groups. Recognition of the importance of group norms in determining individual behaviour is an important departure from neoclassical assumptions about motivation, where only external enforcement (sticks or carrots) affects individual behaviour. In any group, it is easier to generate norms that are broadly consistent with those in society as a whole than to adopt radically different ones, because in the former case the general culture reinforces the group culture.

The community and society[9] within which groups operate affect them in important ways: norms of the community significantly influence the norms that members bring with them to the group and consequently group norms. Moreover, economic and social relationships in society, including property and employment relations, class and ethnic strife, provide the backdrop against which groups are formed, affecting their objectives, power and potential. Thus, analysis of the functioning of particular groups should not look at them in isolation but must include the economic and social context.

A critical aspect of effective group functioning is that the actions of individuals when acting within or on behalf of the group contribute to group aims. Where this does not happen groups become ineffective. For example, if individuals working for a firm shirk, pilfer, recommend rival products or otherwise sabotage operations, the firm is unlikely to be successful. Similarly, cooperatives or public-sector institutions may be used by members/workers to serve their own private interests and as a consequence fail to meet the group objectives. There is not an all or nothing situation, but rather a spectrum ranging from circumstances where members exclusively serve group interests to those where they exclusively serve their own interests. Ways in which individual action is (or is not) reconciled with group objectives is then a vital aspect of group functioning.

The purposes of groups

Groups evolve historically, coming together sometimes for social reasons, sometimes to fight common causes, sometimes to produce (or consume) goods collectively (see Granovetter in Etzioni and Lawrence 1991). Many groups have primarily non-economic functions. Examples are sports clubs and religious organizations. Our focus here is on groups with economic functions, defining 'economic' broadly to include production of goods and services, including non-marketed ones, and activities directed at securing control over economic resources. But it must be noted that the distinction is not clear cut. Not only can most group functions (including social ones) be counted as producing some economic 'output', but also many groups that come together for one (non-economic) reason acquire economic functions and economic groups often spawn 'non-economic' activities.

Group functions may be categorized into two main types, efficiency functions and claims functions, although particular groups may perform both functions. First, efficiency functions correspond to the new institutional economics' view of groups as a response to market failures of various kinds. "Institutions exist to resolve the uncertainties involved in human interaction" (North 1990: 25). A functionalist view of institutions emerges; according to Schotter we can "infer the evolutionary problem that must have existed for the institution as we see it to have developed" (Schotter 1981: 2).

One relevant market imperfection is imperfect information which leads to high transactions costs (see e.g. Nugent 1986). Given indivisibilities of production or consumption and high transactions costs, individuals cannot produce certain goods and services efficiently for themselves. Hence, groups (e.g. firms) are formed to produce goods or to provide common marketing for small-scale producers (e.g. the Kenya Tea Development Authority) or to provide communal facilities (e.g. communal kitchens such as those that developed in Peru during the adjustment crisis) (Graham 1994: 109–110). The externalities associated with non-excludability is another market imperfection which leads to a need for group or collective action to produce public goods. Groups formed to manage common pool resources are an example here (e.g. the Zamjera irrigation schemes in the Philippines). This is the type of group to which theories of collective action apply.[10]

Efficiency functions may be performed by private-sector groups (e.g. firms) or by groups in the public sector (or the community), the first being the probable response to imperfect information and high transactions costs, the latter to the presence of large externalities.

The second important category of functions, neglected in the new institutional economics approach, is claims functions. These occur where a group is formed to advance the claims of its members to power or resources. This is the explicit role of such organizations as associations of the landless, trade unions, cartels and other interest groups. The claims may be advanced against other members of society (e.g. the Farmers' Federation of Thailand sought to advance land claims for the landless against large landowners). Or they may be advanced against the government (e.g. the Sarvodaya Shramadana in Sri Lanka campaigns for a variety of services to be provided by the government). Alternatively, they may aim to enforce legally recognized rights, as for example, helping to ensure that land reforms are fairly implemented.[11]

Some claims functions groups may reduce 'efficiency' in a neoclassical sense. But many groups combine claims and efficiency functions. For example, numerous local organizations both provide public goods and act as a pressure group for their members advancing their claims and their power (Esman and Uphoff 1984: passim).

Mode of operation of groups

Each society and institution has a unique set of laws, rules and norms. But some alternative modes of operation are helpful to differentiate, although any particular group may combine elements of each operative mode. In all cases, the mode of operation helps bring about the coincidence of individual action with group objectives, that, as noted, is essential for effective group action.

Power/control in hierarchical relations and bargaining (P/C). In this mode, one or a group of dominant actors determines, in their own interests, what the rest will do and enforces this by threats of various types, which may be backed up by norms so that the threats rarely need to be used. The basis of the power and the type of threat vary with the institution. The operation of a modern army is a classic example of P/C. Williamson has argued that firms are organized on hierarchical lines, suggesting a P/C relationship, although empirical research indicates that not all firms operate in this way (Williamson 1975, Denison 1990). P/C relations require considerable supervision and monitoring, which can be costly.

Market or quasi-market operations (M). This, of course, is the dominant mode of operation in arms-length transactions between individuals and firms. But it also occurs within groups in the form of a quasi-market, as people are rewarded (and penalized) according to how they contribute to group objectives. Although quasi-markets are rarely the dominant mode of operation, since organizations are formed because of the high transactions costs of market exchange, elements of market incentives are often present within organizations (e.g. piece rates for workers and financial penalties for disobeying group rules). The terms of market transactions within and between groups are conditioned by the power relations prevailing; powerful members of groups and powerful groups can enforce poor terms on weaker elements (see Bowles and Gintis 1993).

Trust/reciprocity (T/R). Here commitment to the group leads to trust, so that actions are carried out in the belief that they will be reciprocated at some time. T/R is more likely to emerge as a dominant mode of operation where relationships among the actors within the group are long-lived (repeated games) and where there is a general ethos of trust, equality and reciprocity. But such relations can also take place between unequal agents (e.g. feudal relations) and among subsets of actors within unequal structures (e.g. among managers in a large firm). Societal norms as well as individual values are relevant to the strength and nature of T/R norms. T/R is the dominant mode in some firms, in cooperatives, in some families and in community organizations.

Tradition/convention (T/C). This mode rests on historically developed values and norms. It is common as the dominant mode in more traditional societies, but also occurs in modern societies, especially within informal organizations. However, it is almost never the exclusive mode of operation. Usually it enforces (and is enforced by) relations of P/C, M or T/R.

Our main focus is therefore on the first three categories.[12] A group often contains a mixture of modes. For example, within a firm there may be T/R and M among managers, P/C and M between managers and workers, while all are influenced by T/C.

The nature of individual motivation is important in determining which mode is likely to be effective and therefore choice of mode. But there is a circular process of causation since where one mode is dominant within a group and even more so within society at large, individual motivation tends to adapt to it. As already noted, the underlying assumption of the neoclassical approach to groups is that of 'rational' individuals whose motive is to maximize their own utility, except within the family. This motivation is perfectly adapted to the M mode but frequently conflicts with a T/R approach. A T/R approach requires either a degree of group solidarity (i.e. some 'I and we' motivation), or that the individual takes a long-sighted view about how individual satisfaction can best be realized.

A contrast between the view of groups being advanced here and that in the neoclassical paradigm can be presented schematically (inevitably in a rather oversimplified way) as in Table 1.[13]

Good and bad groups

The next section contrasts some examples of different types of 'good' and 'bad' groups. 'Good' groups are defined here as those that raise efficiency while generating a satisfactory (interpreted as equitable) distribution of the benefits. Or they may be claims function groups that succeed in improving the position of low-income groups. In contrast, 'bad' groups are defined as those that are dysfunctional (achieve little) or that have benefits monopolized by the richer elements of the group.[14] The aim of our summary of group experiences is to show by example the existence of both good and bad groups and to illustrate important features of each. In practice, as one might expect, it is sometimes difficult to classify groups into 'good' and 'bad'; some are good in some respects but bad in others.

Groups: Some examples

There is a huge area to be covered given the all-embracing definition of groups adopted here. For reasons of space I do not cover the family, private

Table 1 Comparison of Neoclassical and Alternative Views of Groups

	Neoclassical/REM[1]	Alternative
Individual goals	– Exogenous – Maximizing, egotistical	– Evolving, influenced by society and history – Satisficing, commitments to group as well as to self
Norms	– Must be enforced externally, developed to increase efficiency	– Some norms internalized, developed via tradition as well as for efficiency
Group functions	– To enhance efficiency (new institutional economics view) – To constrain trade (rent-seeking view)	– To enhance efficiency – To enforce claims – Historic/evolutionary
Effect of society (e.g. class structure)	– Neutral, except via factor availability and hence prices	– Strongly influences group norms, power, structure
Group characteristics	– Functionalist/efficiency-raising	– Evolutionary/contingent with functionalist elements, increases power and resources of members

Note: 1. REM is 'rational economic man' derived from Folbre.

enterprises or non-governmental organizations, but focus on examples of self-generated community organizations, government-initiated groups and local governments. The examples are drawn from a wide range of secondary sources. They were not randomly collected, but purposively chosen to illustrate the behaviour of different types of group.

Community organizations

Throughout the Third World, large numbers of community organizations have emerged, fulfilling efficiency or claims functions or both.[15] The groups may be organized by the people involved themselves or be initiated by the government or non-governmental organizations. In many cases such groups have failed to meet their objectives. Nonetheless, there have been significantly more successes than would appear probable on the basis of neoclassical assumptions about motivation.

Common resource schemes. It is well known that self-interested motivation leads to the expectation that the commons will be overgrazed. As Hardin expressed it in his classic article:

> Therein is the tragedy. Each man is locked into a system that compels him to increase his herd without limit in a world that is limited. Ruin is the

> destination to which all men rush, each pursuing his own best interest in a society that believes in the freedom of the commons (1968: 1,244).

The basic assumption of the 'tragedy of the commons' hypothesis is that each person acts individually and cooperative arrangements to restrain overgrazing do not develop. This argument has been used to justify the introduction of private property rights.[16]

Olson has advanced a similar scepticism, also from a rational maximizing perspective. He argues that the 'logic of collective action' is such that only if benefits from supporting an arrangement exceed its expected cost to the individual will the person support collective action. This would rarely occur if free-riding was possible.[17]

Yet cooperative arrangements have been initiated in many parts of the world, challenging both sets of arguments. One example is the *Kottapalle Council*, described by Wade. This council was initiated by villagers in a South Indian town of 3,000 inhabitants, with the aim of controlling irrigation and protecting crops against thieves and grazing by other people's animals. The council secures its income from auctioning rights over sheep folding and from the sale of liquor licenses. Its uses its resources to employ common irrigators and field guards who enforce the rules, levying fines when the rules are infringed. The council of nine members is accountable to an annual general meeting of all villagers. According to Wade, the council was highly efficient in devising and enforcing rules that were in the medium- to long-term interest of all the villagers. It managed a severe drought with considerable success, increasing its employment of common irrigators to prevent cheating. The council is autonomous and not part of the official government structures. In a survey of 31 villages in South India, Wade found eight groups with all the features of the Kottapalle Council, eleven with some and twelve with none.

A second example is that of the *Zanjera Irrigation Communities in the Philippines*. Zanjeras date back to the sixteenth century at least. They are "small scale communities of irrigators, who determine their own rules, choose their own officials, guard their own systems and maintain their own canals" (Ostrom 1990: 82). Land is distributed so all have some 'good' (near the source of irrigation) and 'bad' (less well irrigated) land. Members supply their own labour to build and repair dams, canals and other irrigations structures, contributing an average of about two months (unpaid) work a year. It has been suggested that the success of the communities arises in part because of the "strong egalitarian norm, a belief that all persons should contribute to the labour and funds in proportion to the benefit they derive from the irrigation scheme" (Esman and Uphoff 1984, based on Siy 1982). When Zanjera members were asked about problems with the scheme, none complained about the way the water was allocated.

Similar successful autonomous schemes have been recorded concerned with management of common lands in villages in Japan, fishing in Turkey, pastures in Peru and grazing lands and forests in Africa (noted by Ostrom 1990, Runge 1986).

These successful examples that appear to belie the pessimism of Hardin and Olson have several critical features. Firstly, their beneficiaries were not unlimited, but confined to a smallish and stable community, which limited free-riding possibilities while reciprocity and trust had been built up within the community. Secondly, penalties were devised and enforced by the community thereby diminishing the advantages from cheating. Thirdly, interdependence was well recognized. As a consequence of high externalities and relatively small communities, the cost of loss of reputation through cheating was high and community norms became additional sources of enforcement. While government did not play an active role in any of the schemes, its passive support appeared to be important for success. In one failure it was the government's agreement to increase membership against the wishes of the existing members that caused the difficulty.

Cooperatives. In 1975, an report by the United Nations Research Institute for Social Development (UNRISD) concluded, "Rural cooperatives in developing areas today bring little benefit to the mass of the poor inhabitants of these areas... It is the better off rural inhabitants who mainly take advantage of the cooperative services" (UNRISD 1975: ix). Only four out of 14 cooperatives examined in Asia were judged as successful. Similar conclusions were reached by Lele, who identified the causes of failure as lying with 'private interests', which result in "the control of local organizations by the rural political elite ...it frequently results in the perpetuation, if not worsening, of socio-political differentiation" (Lele 1981: 56).

Yet there are many examples of success, such as the *National Dairy Cooperative in India*, which consists of 4,430 village cooperatives encompassing 2 million members. The cooperatives collect milk daily and deliver it to cooperatively owned processing centres. Not only has the cooperative attained a high level of efficiency, minimal corruption and benefited the poor, but it has "also contributed to a weakening of caste and sexual barriers" (Korten 1980: 6).

SANASA (Sri Lanka's Thrift and Credit Cooperative Movement) is another successful example. In 1991 this organization had some 702,000 members in more than 7,000 societies, generating substantial savings and disbursements. Decisions were made at regular participatory meetings. Loans at commercial rates of interest were guaranteed by two fellow members. Over half of the members had incomes below the poverty line, and 70% of borrowers had never approached the formal sector for a loan. There is evidence that the loans provided contributed to a rise in incomes, as well as to reducing borrowers dependency on powerful traders and informal lenders. The repayment record was good, administrative costs were low, and the society had not been co-opted by the elite. Hulme and Montgomery attribute the success of SANASA partly to the strong historical precedents of "symmetrical reciprocal transactions and collective action" (1994: 376) in rural Sri Lanka, which they associate with Buddhist traditions. Moreover, following Uphoff, they argue that expectations of cooperative behaviour, when fulfilled, reinforce that behaviour.

In Santo Domingo, 200 *grupos solidaros* of five to seven bike-riders were formed among the 5,000 *tricicleros* supported by a non-governmental organization to advance loans so that the riders could buy their tricycles. (Previously they had spent one-fifth of their earnings on rent.) Group members guaranteed for each other. The groups subsequently formed an Association of Tricycle Riders which set up a repair shop, a health insurance scheme, and became a pressure group for improving traffic regulations (Hirschman 1984).

Many other examples could be cited. Tentative suggestions for successful community groups can be derived from these examples:

- They tend to be located in fairly unstratified communities with cohesive social structures (but there are exceptions, such as the Indian Dairy Cooperative).
- They tend to have accountable leaders and open procedures.
- They tend to enjoy supportive government.
- They usually adopt a mode of operation within the group as T/R not P/C, and this appears to work best when norms 'outside' are of this type.

On the other hand, there are many sources of failure. Among them are resistance by elites to pro-poor groups or co-option by them, the hostility of governments, male hostility to pro-female groups and internal divisions, often caused by ethnic heterogeneity (see e.g. Esman and Uphoff 1984, Gow et al. 1979). Some groups succeed in some objectives – e.g. improving water use – but still exclude the very disadvantaged and/or women, indicating that the simple classification of groups into 'good' and 'bad' needs modification.

Claims function groups. Many groups have dual functions, both producing various services and acting to advance the claims of the group. Indeed, by being grouped as an entity, people almost automatically become more powerful and therefore better able to advance their claims. Some groups are explicitly formed in order to advance claims. For example, the *Self-Employment Women's Association (SEWA)* in India was formed by a group of women head-loaders, used garment dealers, junk-smiths and vegetable vendors in Ahmedabad, India. The group campaigns for better wages and improved working conditions and defends members against harassment by police and exploitation by middlemen. It also provides training, has developed a savings and credit institution for its members and initiated producer and marketing cooperatives (Cornia, Jolly and Stewart 1987: 215–216). The *Farmers Federation in Thailand* is another example of a claims function group. This farmers' group campaigned to enforce land rights among the landless. Its efforts were unsuccessful, however. Twenty-one leading members were assassinated and eventually the federation collapsed (Ghai and Rahman 1981).

Judgement on campaigning groups depends on the position of their members in society. Where the groups formed are among the privileged (e.g. business cartels) success worsens income distribution. In contrast, groups formed among the weak, both to provide goods and services and to advance their claims, are key to improving their position in society. The effectiveness of claims groups among the disadvantaged depends on the bargaining power they acquire through forming a group. This is a matter of the objective circumstances (e.g. whether there is much unemployed labour in the case of workers' organizations) and of their ability to use their organization effectively (the group's discipline and how confrontational they are prepared to be). The broader socio-economic context, including prevailing norms in society and the nature of economic relations and power, is also relevant as this determines whether attempts at group formation are inhibited by vertical stratification or defeated by confrontational tactics, extending even to physical force. More societal equity in the distribution of assets and income, usually associated with more egalitarian society, tends to make it easier for groups to advance the position of the poor.[18]

Government inspired groups. While according to Esman and Uphoff, "most successful local organizations enjoy the support or at least the acquiescence of government,"[19] the government can stifle local initiative through heavy-handed bureaucracy or deliberate suppression or control of potentially threatening groups. In Thailand, for example, government officials effectively took over the cooperatives; similarly, the government propped up Sri Lanka's *Multipurpose Cooperative Societies* despite their poor standards of service (Uphoff and Wanigaratne 1982). However, some highly successful initiatives originate in central government action but are carried out in and by local communities. For example, the *Taiwan Farmers' Associations* were initiated by the Japanese in 1900 and further developed by the Joint Commission on Rural Reconstruction and the Taiwanese government. The associations were undoubtedly partly responsible for Taiwan's spectacular agricultural growth since 1950. They are private and voluntary but closely follow central government rules. Over two-thirds of farmers are members. The associations receive income from crop and fertilizer sales, and farmers deposit savings with them. They distribute credit and fertilizer, provide extension, run experimental farms and control marketing and the distribution of new seeds, playing a critical role in the introduction of new crops (see Stavis 1974).

Small farmers' groups in Nepal were initiated by the Agricultural Development Bank of Nepal. Small and landless farmers formed groups of 15 to 20 members. Loan requests were approved within the group and advanced against the collateral of the group. Income-generating activities included fish-breeding, piggeries and wheat cultivation. The groups realized a 90% loan recovery rate. Participating farmers increased their income significantly. The groups not only had a significant effect on income generation and social infrastructure (e.g. the construction of latrines), but they increased the sense of self-respect and the bargaining power of the farmers:

> Slowly, but surely all this is increasing the strength of the poorest peasants vis-à-vis the big landowners and money lenders... As they put it in their own words, "Because we are a group now and we stick to each other, we have suddenly become more powerful. The money lenders are afraid to exploit us now. The government officials speak to us, they even speak nicely. We are also no more afraid to enter the bank or the office or the cooperative society" (Bhasin 1978).

Decentralization of government

Local government is an important form of group activity. Decentralization is regarded as a panacea by some, as unequalizing and inefficient by others; some argue that it increases participation, reduces corruption and enhances equity and efficiency. Others say that it increases the power of local elites, transfers power to the inexperienced and extends corruption. The contrasting experiences summarized below indicate that the success of decentralization depends on the context, especially with respect to socio-economic conditions, prevalent norms and the strength of civil society.

Karnataka: An experiment in democratic decentralization. When the Janata Party gained control of the state government of Karnataka in 1983 it introduced quite radical democratic devolution, with elected councils at district (pop. around 28,000) and mandal (a collection of villages or 8,000–10,000 people) levels. Forty percent of state expenditure was devolved to the local level, albeit with fairly strict earmarking and without powers of taxation. The experiment was successful in many respects. It greatly increased participation, raised investment in micro-projects, reduced absenteeism among government employees, increased speed of response and reduced urban bias. The sums involved in corruption are estimated to have decreased and provision of early warning of problems improved. Over 80% of people interviewed expressed satisfaction with the local governments' activities. The majority found the councillors and the bureaucrats honest and helpful. The devolution did not, however, help the vulnerable (scheduled castes and women), who may actually have done less well than before. The Congress Party when it regained power in Karnataka in 1991 brought the experiment in democratic devolution to an end.

Crook and Manor attribute the success of the devolution experiment to "decades of experience with democracy" and to the absence of extreme socio-economic inequalities "which can generate oppression, desperation and vicious conflicts that make it difficult for liberal institutions to function" (Crook and Manor 1994: 50).

Bangladesh. In 1985, General Ershad established councils at the subdistrict level (average pop. 245,000) largely composed of heads of the existing elected Union Councils (pop. around 20,000).[20] The councils were given substantial resources and a fair amount of control over them. Overall, however, the decentralization brought fewer good effects than in the case of Karnataka. Positive impacts included increased participation, raised investment in small-scale infrastructure (but partly at the expense of services such as health care) and improved response time. But corruption was very high (an estimated 30–40% of funds were 'stolen') and the composition of the councils was dominated by the rural elites, who used their power to spend on items that helped them. The poor and vulnerable were almost entirely excluded from benefits.[21] Over 60% of a sample of the population were 'not at all satisfied' with the councils while less than 1% were 'very satisfied'. The subdistrict councils were abolished in 1991, when Ershad fell.

Crook and Manor attribute the weak, corrupt and inegalitarian performance to the nature of rural society in Bangladesh, in particular, to the substantial inequality in land ownership together with considerable landlessness, which contrast with the more equitable society in Karnataka.

North and South Italy. Putnam's account of the effects of the decentralization measures initiated in Italy in 1975 also indicates that these vary depending on social, political and economic conditions. The same measures had much more positive effects in northern Italy than in the south, in terms of both objective indicators of performance, such as numbers of daycare centres constructed and enquiries dealt with, and public views about the adequacy of performance.[22]

Putnam attributes the differences in performance to the existence of a strong civic community in the north. That is, one in which people participate actively in public affairs with "a steady recognition and pursuit of the public good at the expense of purely individual and private ends" (Skinner 1984: 218). "In the civic community...citizens pursue what de Tocqueville termed 'self-interest properly understood', that is self-interest...that is 'enlightened' rather than 'myopic'" (Putnam 1993: 88). In this context, relationships tend to be horizontal, reciprocal and cooperative rather than vertical. Empirical indicators of 'civicness', such as membership of associations, newspaper readership, turnout at referenda, were much higher in northern than southern states.

The source of these differences is traced to political developments in the eleventh century when more participatory republics began to emerge in northern Italy and autocratic government developed in the south. Putnam emphasizes differences in 'social capital'[23] (i.e. 'norms' in the terminology of this paper) as the fundamental source of differences in the quality of government. Less emphasized are parallel differences found in economic structure (more equality less feudal relations) and in the extent of industrialization, which themselves are probably both causes and consequences of the differences in norms.

Some findings and conclusions

This rapid overview of the role of groups in a few areas can be no more than indicative, especially since the examples were not randomly selected, nor purposively designed. The findings discussed below should thus be regarded as hypotheses for future research rather than as well established conclusions.

A general finding is the considerable variation in performance of groups at every level. Our review did not extend to central government, but had it done so it would have revealed a similar variation, with some 'good' governments effectively promoting the well-being of their citizens and others failing even to maintain minimum order. This variation itself indicates that rudimentary explanations of the behaviour of groups are likely to be wrong. There are too many 'good' community groups effectively fulfilling efficiency functions for the views that emerge from a purely short-term self-interested view of motivation to be invariably correct. Moreover, there is enough evidence of the existence of some good governments to suggest that the public-choice view of government is too simplistic; not all governments behave as they would if their actions were purely the outcome of private interests.

For the most part, it seems that among the successful groups observed most adopted a T/R mode of operation rather than a P/C one, whereas the failures were more likely to be associated with a P/C style. This stems in part from the fact that success was defined to include equity as well as efficiency. But T/R type operations appear to be at least as efficient as P/C, and sometimes more efficient, especially where an input of local knowledge improves efficiency (as in irrigation schemes) (Tang 1991, Tang and Ostrom 1993).

There were no examples of market modes within the groups examined, although some groups did use financial penalties as a mechanism of enforcement.

The major reason why the 'public choice'/'tragedy of the commons' views often appear to be wrong derives from the assumptions about motivation, information and trust. If the assumption of strictly selfish motives is replaced by motives that encompass, in part at least, the well-being of others in the community, then more collective action is predicted than in the case of purely 'rational' motivation. Alternatively (or as a complement), if distrust of others is replaced by trust in Prisoner Dilemma type situations, longer term self-interest may lead to cooperation. Motives and trust cannot always be distinguished, as more other-oriented motives tend to be associated with empathy and trust, while where individuals believe in reciprocity, even purely selfish motives may be best served by acting in an apparently altruistic way. These two critical elements – the nature of motives and the degree of trust within groups – do not appear out of nowhere. Rather, they emerge from societal norms and from human experience. This is apparent from the nature of trust: trust, or the absence of it, is forged out of people's prior experience, while motives too are not purely 'innate' but in part socially formed.

There appear to be complex relationships between modes of operation, equity within the group, and equity within the community and society. As some of the examples indicated, groups can be effective in meeting efficiency objectives (e.g. overcoming common resource problems) without helping the poor or vulnerable. This was particularly evident in the case of the Bangladeshi subdistrict councils. However, within-group equity (i.e. a sense of fairness of the distribution of resources) is needed for T/R to be a dominant mode of operation. Moreover, it seems that it is difficult (but not impossible) for an equitable group to prosper in the context of an inequitable community. For example, in the case of the Karnataka government, it seems that socio-economic equity within the community as a whole was an important influence favouring the good functioning of the devolved administration. Hence, there is some connection between the prevalent socio-economic structure and the group behaviour that seems to operate at each level: at the level of central government, good governments are rarely associated with very large inequities in society, among local governments and at community levels. (The business literature suggests a similar connection between mode of operation and equity within firms; see Wilkins and Ouchi 1983, Hansen and Wernefelt 1989.) However, the connection is a loose one – the highly unequal socio-economic conditions in Bangladesh, for example, being associated with successful groups, such as the Grameen Bank and the Bangladesh Rural Advancement Committee as well as the apparent failure of decentralization reported above.[24]

At minimum, passive government support was a feature of the successes, with active government initiation in a number of cases. From this it would seem that groups cannot generally be expected to compensate for failed governments, as they usually require some support from them. Indeed, weak governments are perhaps more likely to provide a breeding ground for conflictual groups, than for socially useful ones. But excessive government intervention can also be harmful, either by weakening or destroying groups where they appear to threaten government power (for example, the Congress Party abolished the Karnataka local governments) or by pushing the groups into a position of dependence.

The general conclusion is that the existence and behaviour of groups cannot be seen in isolation, but are strongly influenced by the society in which they are embedded. In particular, they are influenced by the prevalent norms and the socio-economic structure that they face.

All this is important because groups are crucial for individual and collective well-being, especially among the poor. Positive collective action can avert the tragedy of the commons and advance the situation of the weak, while in its absence, desertification and destitution can go hand in hand. Moreover, bad government can eventually lead to anarchic disaster, while good government can greatly improve social and economic welfare. Contrast, for example, the maternal mortality rate of 14 per 100,000 in Singapore with the rate of 1,000 per 100,000 in the Gambia. Indeed, Putnam argues with some justification that "the character of one's community... is as important as personal circumstances in producing personal happiness" (Putnam 1993: 114), while Esman and Uphoff conclude that " a vigorous network of membership organizations is essential to any serious effort to overcome mass poverty under conditions that are likely to prevail in most developing countries" (Esman and Uphoff 1984: 90).

My main conclusion then is that good group behaviour is an essential aspect of promoting desirable patterns of development, and that to achieve this it is necessary to promote trust and reciprocity at every level of society. Consequently, it is important to identify the conditions likely to generate such trust.

The first important determinant is *history*. This, of course, is a heritage from the past and cannot be changed, although our interpretation of it can change in relevant ways, while today's actions form tomorrow's history. A second condition is the *size and stability of the community*. T/R is more likely to emerge in relatively small and stable communities, so that modernization as such, which brings with it mobility and a growth in the size of relevant communities, may tend to reduce T/R. But offsetting this, modern institutions, such as education, a legal system and government, are designed to increase T/R. Hence the crucial role of good institutions of this type. The third major element influencing norms is the *type of motivation* that is encouraged by the prevailing economic structure, institutions and philosophy.

We are now in a position to explore how group functioning relates to current paradigms of development, in particular to the dominant market paradigm. The evidence presented here suggests that society and community norms, together with socio-economic conditions, are among the most important elements determining group behaviour. The norms that are needed to promote well-behaved groups are those of trust and reciprocity. In addition, conditions of socio-economic equity make T/R more likely and facilitate the development of groups – including governments – that favour the poor. Yet there is an apparent inconsistency between these desiderata and the influence of a laissez-faire market.

The market paradigm – as developed in the textbook version of the neoclassical model of the economy – assumes self-interested individualistic behaviour to be the major engine for generating economic progress, which is orthogonal to the motivation needed for good group behaviour, viz. trust and reciprocity. The effect of emphasizing the dominance and efficiency of 'rational' motives is to encourage people to behave in this way.[25] The more a truly neoclassical style market prevails in every corner of life, the public sector is diminished and quasi-markets are introduced within it, the less T/R type motivation is likely to be present.[26] The market model will thus create a social and institutional 'history' which is unfavourable to good group behaviour.[27] This may be reinforced by rising inequality, often associated with an unregulated market. Hence, there is a danger that where the laissez-faire market becomes too dominant, both norms and socio-economic conditions will move in a way that disfavours good group action.

Ironically, such a tendency would be bad for economic efficiency as well as for people's well-being. Trust in relationships economizes on monitoring, encourages appropriate collective action and permits light-handed government, all of which enhance economic efficiency. The need for trust and teamwork to promote efficiency is well recognized by business itself.[28] Cross-country comparisons show a positive correlation between 'communitarian' values at the country level and growth in GNP and exports (Lodge and Vogel 1987).[29] Similarly, firm profitability has been shown to be positively related to a management 'climate' which emphasizes human resources, as measured by "the employee's perception of how concerned the organization is with his welfare, work conditions, etc." (Hansen and Wiernerfelt 1989: 404). If team-oriented values gain ground within industry, a 'communitarian' market philosophy may help to replace purely self-interested motivation by the 'I and we' necessary for good groups. This has already occurred within the more successful developing countries, such as Taiwan and South Korea where group behaviour and 'I and we' motivation is promoted at all levels of society.

The danger is that elsewhere a crude, individualistic version of the market is being advocated and introduced, with the failure of communism being used to deny the critical importance of groups and collective action in development.

The argument advanced here is not intended to deny the important role of the market in resource allocation and innovation. Rather it is to emphasize that collective action also has a critical role in development. If such action is undermined by the norms promoted in line with an enlarged role for the market and a diminished role for government, groups will still emerge but they will be increasingly more destructive than constructive.

Notes

1. I am among a growing number to appreciate the importance of analysing groups. In this paper I have been much inspired by the work of Bowles and Gintis, Etzioni, Uphoff, Ostrom and Wade in particular. I have also benefited greatly from discussions with Judith Heyer and Rosemary Thorp, from the research assistance of Ruby Gonsen and from comments on a previous draft from Barbara Harriss-White, J. Mohan Rao and Judith Heyer.
2. Within the larger groups (e.g. governments) there are subgroups which in certain areas form the decision-making unit.
3. House et al. (1988) show that that lack of social relationships heightens susceptibility to illness; see also Berelson and Steiner 1964.
4. The UNDP (1995) estimates women's unpaid work as equivalent to 48% of measured income and men's unpaid work as another 22%.
5. See e.g. Gidden (1992).
6. The assumed behaviour of Rational Economic Man in the neoclassical model is described by Folbre as follows: "In the competitive market place where he constantly buys and sells, he is entirely selfish, doesn't care at all about other people's utility. In the home, however, he is entirely altruistic, loves his wife and children as much as his very self" (1994: 18).
7. This is more true of formal than informal groups.
8. Jacques defines 'culture' of a firm as "the customary and traditional way of thinking and of doing things, which is shared to a greater or lesser extent by all its members, and which new members must learn, and at least partially accept, in order to be accepted into service in the firm" (1951: 251).

9. 'Community' refers here to the immediate neighbours, geographically or sometimes metaphorically, while 'society' refers to a broader social environment.
10. See especially Hardin 1968, Olson 1965, Baland and Platteau 1995.
11. Montgomery (1988) suggests that land reform has been more effective where such group enforcement occurred.
12. This categorization of mode of operation is similar to that in some of the business literature, dividing operations of firms into bureaucracies, markets and clans (Wilkins and Ouchi 1983: 471).
13. I am grateful to J. Mohan Rao for suggesting that I summarize the alternatives schematically on the lines presented.
14. Any definition of this kind is unavoidably tendentious and controversial and could be discussed at length. One issue is the definition of 'efficiency' which should include the social desirability of the output of the group (efficient production of e.g. theft by a mafia-type group should not count as 'good').
15. There are numerous studies of individual organizations and also surveys. See the works of Esman and Uphoff, Ostrom, Korten, and Wade.
16. "The only way to avoid the tragedy of the commons in natural resources is to end the common-property system by creating a system of private property rights" (Smith 1981: 467). See also Demsetz (1967). Conversely, some have concluded that strong government action is needed (e.g. Orphuls 1973).
17. "Unless the number of individuals is quite small, or unless there is coercion or some other special device to make individuals act in their common interest, rational self-interested individuals will not act to achieve their common or group interests" (Olson 1965: 2).
18. "Successful local organisation seems more viable where there is relative equity in the ownership and access to productive assets" (Uphoff and Esman 1974: 30). Esman and Uphoff (1984) after analysing the performance of 150 organizations found a positive relation between local community norms and the success of the organizations in assisting the poor either through efficiency or claims functions.
19. Esman and Uphoff (1984: 267), although they find that on balance strong linkages with the government have a negative effect on success.
20. The chair was directly elected, with other councillors chosen from members of the Union Councils.
21. "Decentralisation created significant new openings for village elites to influence government institutions, its overall impact was to intensify already extreme inequalities" (Crook and Manor 1994: 73).
22. Twelve indicators of performance were adopted (Putnam 1993: ch. 3).

23. "Social capital here refers to the features of social organisation, such as trust, norms and networks, that can facilitate the efficiency of society by facilitating coordinated actions" (Putnam 1993: 167).
24. It must be pointed out that in Bangladesh the experiment was too short-lived to come to firm conclusions, devolved powers were limited and the councils were mostly only indirectly elected.
25. It is surely no accident that research has shown that the one exception to the finding that people in society do much less 'free-riding' than economic theory would predict are economists who have learnt about the virtues of Mr Rational Economic Man.
26. Adam Smith himself feared that as a consequence of the advance of the division of labour "all the nobler parts of human character may be in great measure obliterated and extinguished in the great body of people" (Smith 1880: 367). He was referring to the effects of repetitive labour.
27. As Sen (1997) has said, "The purely rational man is close to being a social moron."
28. Some empirical evidence supports the view that "a shared system of beliefs, values, and symbols... has a positive impact on the members' ability to reach consensus and carry out coordinated action" and is more efficient than hierarchical organizations, "participation, communication, creativity and decentralisation" within the firm being positively correlated with growth in sales and profits (Denison 1990: 264–266).
29. In this exercise Japan, Taiwan and Germany are given high scores for 'communitarian values', while the United States, Great Britain and Brazil score low. The exercise is rather arbitrary since it requires ranking the degree of T/R values within firms and nations.

References

Barland, J- M. and J- P. Platteau (1995) *Does Heterogeneity Hinder Collective Action?* Namur: Centre de Recherche en economie du Developpement, Cahiers de la Faculte Economique et Sociales, 146.

Berelson, B. and G. A. Steiner (1964) *Human Behavior: An Inventory of Scientific Findings.* New York: Harcourt, Brace and World.

Bhasin, K. (1978) *Breaking Barriers: A South Asian Experience of Training for Participatory Development.* Hunger Campaign/Action for Development Regional Change Agents' programme. Bangkok: Food and Agriculture Organization of the United Nations.

Bowles, S. and H. Gintis (1993) 'The revenge of Homo economicus: Contested exchange and the revival of political economy', *Journal of Economic Perspectives*, 7: 83–102.

Cornia, G. A., R. Jolly and F. Stewart (1987) *Adjustment with a Human Face.* Oxford: Oxford University Press.

Crook R. and J. Manor (1994) Enhancing participation and institutional performance: Democratic decentralisation in South Asia and West Africa. Report to the Economic and Social Research Fund (ESCOR) January. London: Department of International Development.

Demsetz, H. (1967) 'Towards a theory of property rights', *American Economic Review* (Papers and Proceedings), 57 (2): 347–359.

Denison, D. (1993) 'Organisational culture and human capital', in: A. Etzioni and P. Lawrence (eds), *Socio-economics: Towards a New Synthesis.* Armonk, New York: M. E. Sharpe.

——— (1990) *Corporate Culture and Organisational Effectiveness.* New York: J. Wiley.

Esman, M. and N. Uphoff (1984) *Local Organizations: Intermediaries in Rural Development.* Ithaca: Cornell University Press.

Etzioni, A. (1993) 'Liberals, communitarians and choices', in: A. Etzioni and P. Lawrence (eds), *Socio-economics: Towards a New Synthesis.* Armonk, New York: M. E. Sharpe.

Etzioni, A. and P . Lawrence (1991) *Socio-economics: Towards a New Synthesis.* Armonk, New York: M. E. Sharpe.

Folbre, N. (1994) *Who Pays for the Kids: Gender and the Structures of Constraint.* London: Routledge.

Ghai, D. and A. Rahman (1981) *Rural Poverty and the Small Farmers' Development Programme in Nepal.* Geneva: International Labour Organization.

Giddens, A. (1992) *Sociology.* London: Polity Press.

Gow, D. et al. (1979) *Local Organizations and Rural Development: A Comparative Reappraisal.* Washington, DC: Development Alternatives Inc.

Graham, C. (1994) *Safety Nets, Politics and the Poor.* Washington, DC: Brookings.

Granovetter, M. (1993) 'The social construction of economic institutions', in: A. Etzioni and P. Lawrence (eds), *Socio-economics: Towards a New Synthesis.* Armonk, New York: M. E. Sharpe.

Hansen, G. and B. Wernerfelt (1989) 'Determinants of firm performance: The relative importance of economic and organisational factors', *Strategic Management Journal*, 10: 399–411.

Hardin, G. (1968) 'The tragedy of the commons', *Science*, 162: 1,243–1,248.

Hirschman, A. O. (1984) *Getting Ahead Collectively: Grassroots Experiences in Latin America.* New York: Pergamon.

House, J. S. et al. (1988) 'Social relationships and health', *Sciences*, 37: 11–28.

Hulme, D. (1990) 'Can the Grameen Bank be replicated? Recent experiments in Malaysia, Malawi and Sri Lanka', *Development Policy Review*, 8 (3): 287–300.

Hulme, D. and R. Montgomery (1994) 'Cooperatives, credit and the poor: Private interest, public choice and collective action in Sri Lanka', *Savings and Development*, 3 (18): 359–382.
Jacques, E. (1951) *The Changing Culture of a Factory*. London: Tavistock.
Korten, D. (1980) 'Community organisation and local development: A learning process approach', *Public Administration Review*, 40 (5): 480–511.
Lele, U. (1981) 'Cooperatives and the poor: A comparative perspective', *World Development*, 9: 55–72.
Lodge, G. and E. Vogel (1987) *Ideology and National Competitiveness*. Boston: Harvard Business School Press.
Montgomery, J. D. (1988) *Bureaucrats and People: Grassroots Participation in Third World Development*. Baltimore: Johns Hopkins.
North, D. (1990) *Institutions, Institutional Change and Economic Performance*. Cambridge: Cambridge University Press.
Nugent, J. E. (1986) 'Applications of the theory of transactions costs and collective action to development problems and policy', Paper prepared for the Cornell Conference on 'The role of institutions in economic development', 14–15 November, Ithaca, New York.
Olson, M. (1965) *The Logic of Collective Action*. Cambridge, MA: Harvard University Press.
Orphuls, W. (1973) 'Leviathon or oblivon', in: H. Daley (ed.), *Toward a Steady State Economy*. San Francisco: W. H. Freeman.
Ostrom, E. (1990) *Governing the Commons: The Evolution of Institutions for Collective Action*. New York: Cambridge University Press.
Putnam, R. (1993) *Making Democracy Work*. Princeton: Princeton University Press.
Runge, C. F. (1986) 'Common property and collective action in economic development', *World Development*, 14 (5): 623–636.
Schotter, A. (1981) *Economic Theory and Social Institutions*. Cambridge: Cambridge University Press.
Sen, A. K. (1977) 'Rational fools: A critique of the behavioural foundations of economic theory', *Philosophy and Public Affairs*, 6: 317–344.
Siy, R. (1982) *Community Resource Management: Lessons from the Zanjera*. Quezon City: University of the Philippines.
Skinner, Q. (1984) 'The idea of negative liberty: Philosophical and historical perspectives', in: R. Rorty, J. B. Schneewind and Q.Skinner (eds), *Philosophy in History*. New York: Cambridge University Press.
Smith, A. (1880) *The Wealth of Nations*, Second Edition. Oxford: Clarendon Press.
Smith, R. (1981) 'Resolving the tragedy of the commons by creating private property rights in wildlife', *CATO Journal*, 1 (2): 439–468.
Stavis, B. (1974) 'Rural local governance and agricultural development in Taiwan' reprinted in: N. Uphoff (ed.) (1982-3) *Rural Development and Local Organisation*. New Delhi: Macmillan.
Tang, S. (1991) 'Institutional arrangements and the management of common-pool resources', *Public Administration Review*, 51: 42–51.

Tang S. and E. Ostrom (1993) The governance and management of irrigation systems, an institutional perspective. ODI Irrigation Management Network, Paper 23. London: Overseas Development Institute.

UN (1992) *World Investment Report 1992: Transnational Corporations as Engines of Growth.* New York: United Nations.

UN Development Program (1995) *Human Development Report, 1995.* New York: United Nations.

UN Research Institute for Social Development (1975) *Rural Cooperatives as Agents of Change.* Geneva: UNRISD.

Uphoff, N. and M. Esman (1974) 'Local organisation for rural development: Analysis of Asian experience', reprinted in: N. Uphoff (ed.) (1982–3) *Rural Development and Local Organisation in Asia.* New Delhi: Macmillan.

Uphoff, N. and R. Wanigaratne (1982–3) 'Rural development and local organisation in Sri Lanka', in: N. Uphoff (ed.) *Rural Development and Local Organisation in Asia.* New Delhi: Macmillan.

Wade, R. (1988) *Village Republics: Economic Conditions for Collective Action in South India.* Cambridge: Cambridge University Press.

Wilkins, A. and W. Ouchi (1983) 'Efficient cultures: Exploring the relationship between culture and organisational performance', *Administrative Science Quarterly*, 28: 468–481.

Williamson, O. (1975) *Markets and Hierarchies: Analysis and Antitrust Implications: a Study in the Economics of Internal Organisation.* New York: Free Press.

7 Rethinking Development Assistance: The Implications of Social Citizenship in a Global Economy

E. V. K. FitzGerald

Introduction: The motivations for development assistance

Private charitable initiatives across frontiers are at least two millennia old, and public support of client states is probably even older. Yet 'development assistance' in the institutional form we are familiar with is barely half a century old. It is not surprising therefore that the end of the Cold War has seen a profound reconsideration of the role of development assistance; due in part to doubts about 'aid effectiveness'[1] but mainly because of changing geostrategic priorities, fiscal pressure in donor countries from their own welfare commitments and the shift in perception as to the relative merits of public action and private initiative.

The Development Assistance Committee (DAC) of the OECD, which officially represents the views of donor governments and coordinates their aid efforts, gives three principal motivations for such assistance in the next century (DAC 1996a: 6):

> ...the first motive is *fundamentally humanitarian*. Support for development is a compassionate response to the extreme poverty and human suffering that still afflict one-fifth of the world's population... [They] lack access to clean water and adequate health facilities... sufficient nourishment to live a productive life... basic literacy or numeracy skills. The moral imperative of support for development is self-evident;

> ...the second motive... is *enlightened self interest*... Increased prosperity in the developing countries demonstrably expands markets for the goods and services of the industrialised countries. Increased human security reduces pressure for migration... Political stability and social cohesion diminish the risks of war, terrorism and crime that invariably spill over to other countries;

> ...the third motive is the *solidarity of all people with one another*... one way that people from all nations can work together to address common problems... Sustainable development expands the community of interests

V. FitzGerald (ed.), Social Institutions and Economic Development, 125–142.

> and values necessary to manage... environmental protection, limiting population growth, nuclear non-proliferation, control of illicit drugs, combating epidemic diseases.

Despite this 'moral imperative', the stagnation in official development assistance (ODA)[2] seems intractable. The net real flow has fluctuated around US$ 50 billion a year in 1993 over the past 10 years. Meanwhile the 'burden sharing ratio' has declined: ODA fell from 0.34% of GDP among DAC members in 1983–84 to 0.30% in 1993–94. Within this flow, there have been a number of significant changes since the mid-1980s. The proportion of development assistance channelled through multilateral institutions has risen from a quarter to a third. Japan has emerged as the major bilateral donor. Together, Japan and the members of the European Union now account for three-quarters of all ODA. Further, an increasing share of ODA is allocated to the least-developed countries and to humanitarian emergencies. Non-governmental organizations have taken a growing role in the delivery of assistance – particularly in conditions of extreme poverty and in conflict situations. Last but not least, official development finance has declined from two-thirds to one-third of total net resource flows to developing countries, due to the extraordinary expansion of private investment flows towards 'emerging markets'. Nonetheless, ODA remains crucial to the low-income countries, accounting for as much as 11% of GDP in the case of sub-Saharan Africa.

The emergence of an integrated 'global' economy clearly presents a profound problem for the form and organization of development assistance. On the one hand, a single world market for capital and commodities presents new opportunities for the dynamic developing countries to modernize their economies on the basis of foreign financial resources, technology transfer and access to vast consumer markets (World Bank 1995). From this point of view, the task of aid agencies should be to help poor countries take advantage of these opportunities and to make sure that the emerging institutions of global governance such as the World Trade Organization ensure equitable treatment to all participants in the global market. Aid should also support public action to tackle global cross-border externalities such as environmental degradation and the drugs trade. These efforts to underpin a global market correspond to the second and third of the three motivations as defined by the DAC mentioned above.

On the other hand, the poorest countries and vulnerable groups that do not possess the resources to compete effectively have become more vulnerable to exogenous shocks and are falling further behind in the race for technological competence (UNRISD 1995). From this point of view, aid agencies should become part of an international social safety net which reflects not only the global ethical responsibilities of the rich for the poor, but also the claim of the poor upon the rich as members of the same global community. This corresponds to the first of the three DAC motivations, but the international institutions that reflect this claim do not have a clear place in global economic governance.

In this paper I address these two very different models of development assistance from the viewpoint of the problematic nature of citizenship within a global economy without a global state. First, I explore the model of aid derived from the notion of 'incomplete global markets'. This appears to be the logic of the approach espoused by the World Bank and the International Monetary Fund, where the central role of development assistance is to support the full integration of developing countries into the global economy by correcting for failures in product and capital markets. I also argue that because free migration – and thus truly global labour markets – are not part of this model, the logical implication is that citizenship itself becomes an exclusive form of property and thus an economic asset.

Second, I discuss an alternative model of aid based on 'global entitlements'. The right to the satisfaction of certain universal entitlements to livelihood is contained in core United Nations agreements among governments as well as in international law. Here the central role of development assistance is to ensure the provision of minimum levels of health, education and nutrition, as reflected in specific poverty and life expectancy targets. I argue that the recognition of this claim logically implies a concept of global social citizenship, and thus by extension, a system of international taxation to support it.

I conclude this paper with a brief discussion of the ways in which changes in the global political economy may affect the evolution of development assistance in the future.

The 'incomplete markets' model: Development assistance as a correction for market failure

The 'developing countries', lately joined by 'transition economies', are now central to the world economy as well as containing the greater part of the world population. As Table 1 indicates, the 'rest of the world' other than the industrial countries now accounts for nearly half of global production and more than half of global investment, even though its share of world trade is somewhat less. Capital flows to the 'rest of the world' from the industrial countries now exceed US$ 150 billion a year as international capital markets have been privatized and integrated worldwide; although as they only represent 5% of fixed investment, local savings still provide the main source of development finance. Indeed, the greater part of resource transfers from the developed to the developing countries now takes non-official 'private' forms,

Table 1 The Structure of Global Capital Accumulation, 1994

	No.	% World GDP	% World Exports	% World Savings	% World Invest.	S/Y	I/Y	(S–I)/Y
World	183	100.0	100.0	100.0	100.0	23.1	23.7	–0.6
Industrial economies	23	54.6	70.2	47.0	45.8	19.9	20.0	–0.1
USA	1	21.2	13.2	14.3	14.0	15.6	17.4	–1.8
Japan	1	8.4	8.4	11.5	11.2	31.7	28.8	2.9
EU	15	21.1	40.4	20.1	19.6	22.0	19.0	3.0
Rest of World	160	45.4	29.8	53.0	54.2	26.6	27.9	–1.3
Net creditors	7	2.1	3.9	1.9	2.0	20.8	22.2	–1.4
Net debtors	125	38.0	22.2	46.6	47.4	28.0	29.2	–1.2
Market borr.	23	23.0	16.5	33.5	33.2	33.3	33.8	–0.5
Division borr.	33	10.3	4.0	10.0	10.3	n.a.	n.a.	n.a.
Off borr.	69	4.8	1.7	3.0	3.9	14.4	18.9	–4.5
Transition economies	28	5.3	3.7	4.4	4.5	19.0	20.0	–1.0
Devloping countries	132	40.1	26.1	48.5	49.5	27.6	28.9	–1.3
Africa	50	3.3	1.7	2.5	3.1	17.3	21.7	–4.4
Asia	30	23.1	16.6	34.4	34.1	34.0	34.5	–0.5
Middle East and East	18	4.8	3.9	4.4	4.3	21.1	20.8	0.3
Western hemisphere	34	8.9	3.9	7.1	8.1	18.2	21.4	–3.2
Balanced account								
World	183	100.0	100.0	100.0	100.0	23.6	23.6	0.0
Industrial economies	23	54.6	70.2	46.3	43.6	20.0	18.8	1.2
Rest of World	160	45.4	29.8	53.7	56.4	26.6	27.9	–1.3

Note: The country group shares in world investment and savings were estimated by weighting the respective rates for each country grouping by their share in world GDP. As the IMF estimates for world investment and savings are not in balance (as they must be by definition), 'balanced accounts' are also shown where the error (some 0.6 % of world GDP, mainly attributable to unreported investment income) is allocated to industrial economies' income and savings, and then the resulting shares are recalculated.
Source: IMF (1996) for shares of world GDP (on a ppp basis) and exports of goods and services, and for savings/investment rates (i.e. shares of current GDP) by country groupings.

as Table 2 indicates. What is more, the observed rates of interest paid by developing countries on their debt do not differ significantly from the higher levels paid by public borrowers within the OECD itself.[3]

So is official development assistance needed at all? In principle, all the tasks carried out by aid could be done by the private sector. Finance for economic and social infrastructure projects could be provided by international equity or bond markets; balance of payments support could be provided by commercial banks; and less-developed country governments could borrow to fund welfare expenditure even in emergencies – just as developed counties do.

Table 2 Pattern of Development Assistance, 1994

Net Resource Flows to Developing Countries (US$ billions)	
Official development assistance (ODA)	59.7
Other official development finance	10.5
Total official development finance (ODF)	70.2
Export credits	3.2
Direct foreign investment	47.0
International bank lending	21.0
International bond lending	32.7
Other market flows	4.0
Grants by non-governmental organizations	5.7
Total private flows (from DAC members)	110.4
Total net resource flows	183.8
Distribution of ODA by Income Group of Recipient (per cent)	
Least-developed countries	24.2
Other low-income economies (LICs)	14.7
Total LICs (per capita GNP < $675 in 1992)	38.9
Lower middle-income countries ($676–$2,695)	20.8
Upper middle-income countries ($2,696–$8,355)	3.8
Total middle-income countries	24.2
High-income developing countries	2.9
Other (geographically unallocated)	34.0
Total ODA	100.0
Major Uses of Aid (percentage)	
Social and administrative infrastructure	27.3
of which, health and education	14.0
Economic infrastructure	27.5
of which, energy	11.6
Agriculture	8.6
Industry and other production	6.1
Food aid	1.9
Programme assistance	8.9
Other	19.6
Total	100.0

Source: DAC (1996b).

Technical assistance could be provided by consultancy firms and, in the last resort, international charities could support the poorest. Nonetheless, capital flows to low-income countries mainly take the form of official flows ('aid'), which allow a current account deficit in the balance of payments to be maintained, and investment rates to be kept higher than savings rates. These flows represent almost 5% of the income of official borrowers (see Table 1) and thus one-quarter of their investment. At the same time, however, these aid flows represent as little as 0.2% of world income, and thus less than 1% of global savings; or approximately 0.4% of industrial countries' income and 2% of their savings.

In this model, the need for aid arises because there is 'market failure' in the strict sense of their existing a wide range of activities in developing countries which are economically profitable but which the private sector (domestic or foreign) is unable or unwilling to finance because the external benefits cannot be captured by the investor or because the risks involved are too great. Further, there is a range of social services – such as health and education – that have a positive return to the economy as a whole in the long run, but which governments cannot fund due to fiscal constraints. In other words, ODA is seen as complementary to, rather than as a substitute for, private investment.

We can identify five causes of market failure in this sense. First is *asymmetric information*, where private investors have insufficient information about profitable projects, but where an international body can act as an effective intermediary due to its analytical capabilities or by increasing the information available can reduce the cost to investors. Second is *economies of scale, scope and learning*, where the scale of the investment required or the timescale involved is too great for individual foreign investors. Here again an international public body can act as an effective financial intermediary by relying on the long-term borrowing capacity of member governments. Third is the *significant externalities* which occur where the economic benefits of (say) education and health provision would not accrue to the individual investor, although they clearly increase profits of investors as a whole in terms of future production costs, market expansion and technological innovation. Fourth is *agency problems*, where there exist serious risks of non-performance by debtors or macroeconomic instability which reduces asset values and where an international agency can reduce the risk through policy conditionality and policing of 'free-riding', and in the last resort control outbreaks of violence that threaten global markets as a whole. Fifth is *coordination failure*, where joint action by public or private agents would be to the benefit of all – typically in spheres such as environmental maintenance, financial regulation, crime prevention and disease control – but where an individual agent cannot ensure such joint action and may benefit more from individualistic behaviour.

These problems are of course systemic to all markets, particularly financial markets. But they are especially strong in international capital markets. Development cooperation under these circumstances would in effect be designed to overcome market failure; thus to the extent that markets work better it could be gradually reduced. In consequence, countries' eligibility for aid would be based on incomplete market access rather than a particular level of per capita income. It would be possible to both 'graduate' most middle-income countries and even some of the larger low-income countries from eligibility *and* retain the possibility of temporary access in emergency situations when market access is interrupted.

The justification of aid in this model would be that by making international markets work more efficiently, the welfare of both donors *and* recipients is increased. Donor countries get a higher return on investment because risks are reduced; while the level of investment and growth in recipient countries is also raised. In terms of political theory, this approach would correspond to liberal arguments for residual state involvement in economic and social affairs within a single sovereign jurisdiction. It would be quite distinct from the traditional utilitarian argument for aid as the redistribution of income in order to increase global welfare on the basis that the marginal utility of income for donors is less than that of recipients because the per capita income of the former is higher than that of the latter. In terms of moral philosophy, this approach to aid would appear to correspond to commutative ('horizontal') equity and thus ultimately to a sense of 'fairness' as the basis for a liberal theory of justice (Rawls 1972).

Examples of this picture of aid are not difficult to find and clearly represent a strengthening trend. The role of the International Monetary Fund in policy-conditioned balance of payments lending to developing countries is well established, and the recent loans to Mexico[4] underline the international financial institutions bailing out of private investors from rich countries. The World Bank increasingly sees its role as to support private investment, and it would wish to expand the activities of the International Finance Corporation. The Asian Development Bank emphasizes its role as underpinning large infrastructure projects where private funding predominates. The Inter-American Development Bank actively supports the privatization of utilities and charging systems for health and education. Most OECD members provide some sort of export and investment risk insurance for foreign investors in developing countries. The Club of Paris attempts to write off old debts in order to restore the credit-worthiness of the least-developed countries.

The gradual process of reform of international trading arrangements – particularly the WTO – can be read in much the same manner. The gradual dismantling of import barriers by industrial countries is intended to benefit developing-country exporters as well as developed-country consumers, and should increase production efficiency on both sides. Similar arguments are made for commodity stabilization schemes of various kinds, which reduce risk and increase the welfare of both producers and consumers. The rather more controversial proposals for the inclusion of labour standards in future global trading arrangements are explicitly seen by their proponents[5] as a means to help the poor within developing countries, defending them from their own governments and entrepreneurs in the name of horizontal equity. The justification of similar trade measures for environmental protection, in contrast, could be expressed in terms of the externalities arising from private investment and would thus justify public action in the way discussed above.

At first sight, this argument might seem to exclude humanitarian emergencies or situations of extreme poverty. Nonetheless, the logic of market failure might be extended to these cases too. Emergency relief can be seen as a consumption loan over the lifecycle until recipients can become self-supporting (on a parallel with 'workfare') or as the cost of maintaining security in a particular region so that economic institutions do not break down in a wider area. Clearly, however, this model of aid is more appropriate to middle-income developing countries where the private sector (ranging from large oligopolistic corporations to small farmers and artisans) is strong enough to with adequate support take advantage of international markets.

A more general problem with this approach derives from the so-called 'Pareto conditions' for a free-market equilibrium to lead to welfare maximization. It can be shown that once various sources of market failure, such as those discussed above, have been overcome, a perfectly free market in equilibrium maximizes welfare for a given prior distribution of assets – whether financial wealth or human capital – by equating the marginal utility of consumers with the marginal cost of production in all markets. However, there is a distinct equilibrium for each such prior distribution of assets and no reason to believe that over time the market will lead to that distribution changing in a welfare-enhancing manner.

For the poor in developing countries one of the key characteristics of this prior distribution is precisely their location in those countries, in other words, their citizenship. The emerging global economy evidently implies integrated and minimally regulated international markets for goods and capital, but not for labour. The traditional economists' response to this problem is based on the Stolper-Samuelson theorem: the potential economic benefits of migration in terms of increased employment and higher wages can be achieved just as well by labour-intensive exports and international investment in developing countries. In other words, free markets and capital mobility substitute for labour mobility. There is thus no need to liberalize international labour markets. However, it is clear that people in developing and developed countries do not share this view, because the former are continually trying to migrate and the latter to prevent them from doing so. Indeed one of the major justifications for European Union aid to its poorer neighbours is to raise the level of employment there in order to reduce the pressure for illegal immigration.

In effect citizenship[6] is the single most important asset most people possess in developed countries: it is a better explanator of differences between their incomes and those of people in developing countries than skills or labour productivity. In this sense citizenship represents a claim on the accumulated social capital of the relevant country, including its power in international economic arrangements. In the specific case of the poor, of course, citizenship represents a claim on welfare payments which, discounted to the present, can be understood as a capital sum.

At first it might seem that citizenship cannot be properly considered as an economic asset as it has no market. However, there is an implicit market in citizenship in the sense that many developed countries – and small developing countries hoping to attain status as financial centres – confer work permits and eventual citizenship on persons who are highly skilled or are substantial investors. Admittedly such citizenship is inalienable, but it can often be conferred on spouses and usually on children. Moreover, many forms of property (particularly land) are restricted in their sale and may, for instance, revert to the state if the owner no longer uses them, as in a sense happens when a citizen dies.

A recent libertarian proposal in the Von Hayek tradition which would in effect formalize the 'purchase' of social citizenship is that immigration should be permitted if the applicant can provide sufficient capital so the net per capita wealth of the host country is not reduced.[7] This is rather more consistent with actual practice, as well as with the tenets of neoclassical trade theory, although it poses the obvious problem that the remaining population in the country of origin will then have a *lower* per capita wealth and presumably a net welfare loss.

In sum, one of the major difficulties inherent in the emerging global economy – and by extension, in the market-failure approach to aid – is the effective prohibition of international labour mobility which, as we have seen, makes citizenship an economic asset. Justification of aid in terms of horizontal equity (or commutative justice) does not really meet the requirements of the liberal political theory. The contractarian approach to economic justice proposed by Rawls involves the application of a test based on the 'veil of ignorance' in order to assess the fairness of a particular set of institutional arrangements (Rawls 1972). In this case, the appropriate test would presumably be whether a rational person would be willing to be born into the world irrespective of citizenship. Clearly the answer is negative, in which case it would appear that the present arrangements can be properly judged as unjust (Beitz 1979).

The 'human entitlement' model: Development assistance as global social citizenship

Consideration of 'vertical equity' as a model for aid, as opposed to the 'horizontal equity' of the previous model, implies a quite different approach based on the entitlement of the poor or vulnerable to economic support on the grounds of a common humanity – recognizing their membership of a global moral community similar to that existing within a single country.

The ethical or moral basis for aid, despite the apparent status of the first DAC motivation quoted above as a 'moral imperative', is not unambiguous. The argument for aid on grounds of humanity derives from the obligation to relieve human suffering when this can be done at little personal cost, which is a universal obligation in relation to all humans simply by virtue of our shared humanity. It is based on the Kantian proposition that the principle of respect for persons is a minimal condition for moral relations to exist at all (Benn 1988). This means that the moral foundations of aid are derived from the

> obligations of humanity (that) require resources to be transferred to the poor, irrespective of state or national boundaries. These obligations rest on individuals, though they may be satisfied through the agency of the state, and they require wealthy individuals to transfer sufficient resources to provide the poor with the means of survival (Opeskin 1996: 22).[8]

It is difficult, however, to establish any clear ethical argument for aid beyond this basic 'human entitlement'.[9] The Aristotelian notion of redistributive justice is usually applied (as in the contractarian theory of Rawls) to individuals within an identifiable community. Yet to apply it internationally poses two problems: first whether the contractarian responsibility of individuals extends beyond state boundaries and second whether states can properly be considered as moral agents in the international sphere (ibid.: 28–30). It can be argued that even within the liberal scheme, principles of justice are relevant, because international trade and investment activities produce substantial aggregate benefits, while participation is largely non-voluntary for poor countries (Beitz 1979). Equally, it can be argued that these activities do not constitute a relevant scheme of cooperation at all – particularly for individuals upon whom moral obligations ultimately rest – as is evidenced by the nature of the international organizations that exist in practice (Barry 1973).

Despite these philosophical ambiguities, there does appear to exist a significant, and growing, obligation on states under international law to confer development assistance. Traditionally, international law has been concerned with the international rights and duties of states. However "one of the most significant advances of international law has been the development of rules and principles governing the rights and obligations of individuals" and in particular the rules of international law offering protection to individuals under the law of human rights (Hill 1994: 275).[10] This represents a significant departure from the positivist approach to international law based on custom and treaty, and "in practice, the bulk of human rights law operates to prevent the state from causing 'harm' to its own nationals. This does not mean merely physical injury, but can encompass economic, social, legal or intellectual harassment" (ibid.: 276–277). The source, of course, is the Universal Declaration of Human rights adopted by the UN General Assembly in 1948. The effectiveness of these rights depends on the existence of mechanisms for implementation, which at present is mainly limited to the 'first generation' civil and political rights that form the core of most human rights treaty regimes.

UN resolutions, from chapters IX and X of the UN Charter itself through to the 1986 Declaration on the Right to Development, recognize the urgent need to deal with economic and social problems. Indeed, under international public law "there is probably also a collective duty of member states to take responsibility to create reasonable living standards both for their own peoples and for those of other states" (Brownlie 1990: 259). However, these 'third generation' human rights are commonly assumed to be 'emerging' rather than an established part of international law as the first generation rights are, although the right to development is prominent in both literature and diplomacy, while the right to an adequate standard of living (otherwise the 'right to food') has received substantial recognition as a legal standard (Brownlie 1987).

This recognition of international obligation clearly cannot be interpreted as reflecting 'global citizenship' in the full sense, because nothing approaching a 'global state' exists.[11] The original idea of 'civil' citizenship was one of equality before the law and active participation in classical society, which clearly excluded foreigners as well as slaves and women. Even the universalist ideal, culminating in the French Revolution, established a 'political' citizenship based on the theory of consent which only extended to (male) members of a single nation. Moreover, the contractarian approach to citizenship explicitly rejects the communitarian notion of the 'common good' by which global public action can be justified. Indeed, it can be argued that this tension is inherent in capitalism as a social formation where "the tension between... the egalitarian promise of citizenship and the inequalities of the market economy have surfaced and been negotiated," the result being the welfare state (Riley 1992: 187).

Nonetheless, in a sense the practice of both international public law and international development assistance, despite their evident shortcomings, can be interpreted as reflecting a degree of implicit recognition of 'global social citizenship' in terms of both rights and duties. Marshall's notion of social citizenship (also termed, interestingly, the 'citizenship of entitlement') consists of "the whole range from the right to a modicum of economic welfare and security to the right to share to the full in the social heritage and to live the life of a civilized being according to the standards prevailing in society" (Marshall 1950: 11). This is not, therefore, an argument from compassion, which focuses on the point of view of the donor as (global) citizen and where the recipients are not perceived as citizens with entitlements to benefits and rights of participation in decisions which affect them, but rather as beneficiaries of largesse. A depersonalized relationship based on entitlement is essential if recipients of social benefits are to be citizens rather than subjects (Oliver and Heater 1994). Social citizenship also depersonalizes the function of giving, converting it from a voluntary act by a few 'good citizens' to a duty of all citizens who can afford to do so to pay tax so that the needs of the body of the citizenry are met.

The global fiscal revenue required to meet even a minimal obligation to render humanitarian development assistance is much greater than the present level of aid. If we adopt as a minimal income requirement the World Bank's standard of one US dollar per diem, then 1.3 billion people currently fall below this standard (DAC 1996a: 6). The annual transfer of income required to bring this one-fifth of the world's population up to the minimal level would be some US$ 250 billion per annum, more than four times the current level of development assistance.[12] In terms of material objectives, the DAC has accepted the following specific targets for all developing countries by 2015 (DAC 1996a: 9–11): (i) reduction by half of the proportion of people living in poverty, (ii) universal primary education, (iii) reduction of infant mortality by two-thirds and maternal mortality by three-fourths, (iv) universal access to primary health care and family planning methods and (v) reversal of current trends in the loss of environmental resources. These goals have not been costed and they would involve more domestic than external resources, but they can be assumed to raise the presumptive resource requirement to at least 2% of the GDP of the industrial countries, even if all other aid objectives were abandoned.[13]

Social citizenship implies not only entitlements to welfare payments. It also entails the obligation of better-off citizens to pay tax in order to finance assistance to the poor if the system is to be something more than social insurance, which would merely be an extension of the first model of aid. Such taxation is implicit in the existing funding of ODA out of tax revenue. A global progressive income or wealth tax in addition to existing taxation would be the ideal theoretical solution, but is hardly practicable.[14] An alternative approach would be to base such a tax on 'global' activities, specifically on the income derived from transactions arising from the process of globalization itself. One possibility is a duty on international airline tickets,[15] as this would be relatively easy to collect (through the International Air Transport Association) and would fall on the better-off. Another is the so-called 'Tobin tax' on short-term capital flows,[16] although this would be extremely difficult to administer.

The political legitimization of such a tax would also require a system of distribution of development assistance that effectively reached the poor and thus an even greater degree of aid conditionality than is the case at present. This might seem to involve a considerable reduction of sovereignty by recipient states, but there are two reasons for believing that this would not be the case. First, the identification of specific targets in terms of (say) primary health and education services would allow a contractual relationship to be worked out similar to that which exists between central and local government in federal states (e.g. the United States or Germany) or within the European Union. Second, the recognition of universal rights and obligations would radically change the present asymmetric relationship between aid 'donors' and aid 'recipients', much as the replacement of charity by welfare systems did in Western Europe.

Conclusion: Development assistance and international political economy

In this paper I have discussed two conceptual approaches to the future of development assistance in a global economy. The first is based on an apparently contractarian approach which emphasizes the role of aid in overcoming international market failure and thus ensuring horizontal equity. Nonetheless, the logic of this approach seems to lead to citizenship of a developed state becoming a significant economic asset, and the resulting situation clearly fails the standard test for 'fairness' as the basis for a liberal theory of justice. The second approach is based on a more communitarian notion of social citizenship on a global scale. Emphasized here is the role of aid in guaranteeing livelihood for all human beings on the basis of recognition of their humanity, thus ensuring a minimum of vertical equity. However, this argument seems to lead to financial obligations on the part of developed countries which can be supported only by an implicit or explicit system of global taxation and some international system to supervise expenditure.

The two models are not entirely antithetical, if only because in an ideal world both vertical and horizontal equity would be attained. Indeed they might be thought of as complementary to the extent that the minimal provisions of international social citizenship would make the residual inequalities arising from individual state citizenship easier to justify. Moreover, in practice there is a clear tendency at present for the horizontal equity model to be used in relationships with middle-income developing countries (e.g. in Latin America) and the vertical equity model to be used in relation to low-income countries (e.g. in Africa). Yet this may be a false distinction. On the one hand, any entitlements derived from global social citizenship must be universal; they cannot be confined to a particular (albeit poor) group without becoming a modern form of charity. On the other hand, low-income countries need effective integration into world markets as much as any other countries do, and probably more so since many of their problems stem from their vulnerability to fluctuations in commodity prices.

In practice, there is every reason to believe that fiscal constraints and changing geostrategic priorities will continue to constrain absolute levels of development assistance. Despite the decline in North American aid commitments, the United States appears to maintain a high degree of doctrinal hegemony in international economic institutions. In contrast, Japan and the European Union – now the major aid donors – do not appear to have developed a clear strategy for global economic governance. Under these circumstances even maintaining present levels of aid flows in real terms might be considered optimistic, although this would imply a steady decline in the 'burden sharing ratio' to 0.15% by 2020. In sum, it appears likely that the emerging system of global governance will privilege the international finance and trade institutions committed to the first 'horizontal equity' model discussed in this paper, despite its implications for citizenship.

Nonetheless, there is some evidence that changes in the international political economy which respond to distinct institutional forces may eventually tip the balance back towards the second model. *First*, the multilateralization of global economic governance means that systems of intervention have to be based on rules rather than on the discretion of hegemonic powers, as the consolidation of the WTO shows. In consequence, poor countries have a greater opportunity to press for the rationalization of development assistance. *Second,* recent experiences with humanitarian intervention have undoubtedly been based on concerns for geopolitical stability. But the explicit recognition of the rights of vulnerable groups that this involves, independent of the position of states, is creating a precedent for global social citizenship, that is, a humanitarian entitlement to aid. *Third*, the emergence of the European Union as a regional political-economic grouping which can enforce social provisions appears to involve for the first time the creation of trans-state social citizenship; while the gradual international harmonization of taxation offers the possibility that expenditure provisions might also be standardized.

Moreover, the treatment of commercial concerns in international law is developing rather more rapidly than the treatment of poverty. Generally, when an individual or company enters into a contract with a state, that contract is governed by the municipal law of one of the parties, usually the state.

This parallels the practice with contracts between firms from different countries, where the municipal jurisdiction of one of the contracting parties – or a third location, such as New York – is chosen (Hill 1994). Nonetheless, there is a growing tendency for contractual relations between a foreign national and a state to give rise to international responsibility for that state, and thus require international arbitration. Similarly, the gradual adoption of the 1964 International Convention for the Settlement of Investment Disputes provides a formal mechanism for the settlement of investment disputes between contracting states and nationals of contracting states, subject to prior consent – bringing a large measure of stability and certainty to foreign investment matters (Dixon 1993).

These arrangements, of course, correspond to the global effort to improve horizontal equity and reduce market failure in the first model discussed above. They may, however, hold positive implications for the development of global social citizenship. In particular, an increasing proportion of the expatriate employees of multinational firms either are not citizens of an internationally powerful state even if the mother company is registered there or they are employed by a multinational firm originating in a developing country. In either case, there is no clear means of guaranteeing the civil rights of such employees unless some kind of universal citizenship is implicitly recognized. What is more, the growth of non-governmental organizations (as opposed to intergovernmental bodies) has led to their acceptance as bearers of rights and duties under international law (Detter 1994: 125) even though the status of their employees – particularly those of the smaller aid agencies – overseas remains unclear. Both these trends should reinforce the pressure for effective enforcement of 'first generation' rights on a global basis for temporary residents, which in turn would support the case for global social citizenship in a wider sense and thus for a redefinition of development assistance.

In an ideal world, a global entitlement to development assistance based on a concept of global citizenship would be supervised by the United Nations itself, possibly through a 'social and economic security council' (Mahbub ul Haq 1996). As there is little prospect of this in the foreseeable future, it may well be better in practice to secure the inclusion of such entitlements in the emerging governance institutions of the global economy itself, such as the WTO and the other eventual successors to the Bretton Woods institutions (FitzGerald 1996).

Notes

1. Although there are clearly serious problems of aid effectiveness (lack of donor coordination, failure to support local institutions, local corruption, perverse macroeconomic consequences, insufficient government commitment to agreed programmes, inappropriate conditionality etc.), these are not the topic of this paper. For a balanced discussion, see Cassen and Associates (1994).
2. Data in this paragraph is drawn from DAC (1996b); see also Table 2.
3. Data in IMF (1996) suggests that the rate of interest effectively paid on debt (defined as debt service divided by debt outstanding) during 1993–95 in Africa was 12%, which refers mainly to official debts; while the average rate for all developing countries 'with debt service difficulties' was 11%, not significantly different from the yield on government debt in southern Europe.
4. The US$ 50 billion support operation in 1995 was almost as large as total ODA recorded by the DAC for that year.
5. The motives of some of the supporters of these proposals, such as those related to labour-intensive industries in developed countries, may be suspect, but this does not detract from the merits of the original proposal.
6. I use the term in preference to 'nationality' because one state can contain many nationalities.
7. "Free immigration for people who can prove to be self-reliant (privately insured for sickness and old age), not being a risk for the immigrant country's social security system" (Giersch 1996: 4).
8. Opeskin effectively dismisses the 'consequentialist' argument against aid (that it is largely ineffective and may even do harm) by pointing out that this is a pressing moral argument for reforming aid institutions, not for refusing aid as such.
9. In Sen's terminology, these would be 'social' entitlements in contrast to market-based entitlements based on income or wealth (see Dreze and Sen 1989: ch. 2).
10. The other aspect is personal responsibility for piracy, crimes against humanity, etc.
11. The UN General Assembly cannot really be considered as anything more than an intergovernmental association.
12. This is a conservative estimate based on the assumption that this group is evenly distributed about the mean (i.e. $0.50 per diem) and that the transfer would be completely efficient (i.e. no administrative cost) so that the mean is raised to one dollar.
13. At present, only 15% of ODA is allocated to health and education and of this only a small proportion goes primary provision (see Table 2).

14. Whether existing aid funding is progressive in the distributive sense depends upon what the marginal source of fiscal income for aid donors is considered to be. At the margin, OECD countries under public-sector borrowing constraints and reluctant to increase tax rates might well reduce their own social expenditure, which is hardly a desirable outcome!
15. Originally proposed in order to finance the United Nations itself, which has inevitably made it somewhat unattractive.
16. Originally proposed in order to reduce capital account and exchange rate volatility, although it would not in fact have any appreciable effect on speculation at feasible rates.

References

Barry, B. (1983) *The Liberal Theory of Justice.* Oxford: Clarendon Press.

Beitz, C. (1979) *Political Theory and International Relations.* Princeton, NJ: Princeton University Press.

Benn, S. I. (1988) *A Theory of Freedom.* Cambridge: Cambridge University Press.

Brownlie, I. (1987) *The Right to Food.* London: Commonwealth Secretariat.

——— (1990) *Principles of Public International Law* (4th edn). Oxford: Clarendon Press.

Cassen, R. H. and associates (1994) *Does Aid Work?* (2nd edn). Oxford: Oxford University Press.

DAC (1996a) *Shaping the 21st Century: The Contribution of Development Co-operation.* Paris: OECD Development Assistance Committee.

——— (1996b) *Development Cooperation Report, 1995.* Paris: OECD Development Assistance Committee.

Detter, I. (1994) *The International Legal Order.* Aldershot: Dartmouth.

Dixon, M. (1993) *Textbook on International Law* (2nd edn). London: Blackstone Press.

Dreze, J. and A. K. Sen (1989) *Hunger and Public Action.* Oxford: Clarendon Press.

FitzGerald, E. V. K. (1996) Intervention versus regulation: The role of the IMF in crisis prevention and management. UNCTAD Review. Geneva: UN Conference on Trade and Development.

Giersch, H. (1996) 'Rules for faster growth in the world economy', International High-Level Experts Meeting on Globalization and Linkages to 2020. Paris: Organization for Economic Co-operation and Development (In Mimeo).

Hill, J. (1994) *The Law Relating to International Commercial Disputes.* London: Lloyds of London Press.

IMF (1996) *World Economic Outlook.* Washington, DC: International Monetary Fund.

Mahbub ul Haq (1996) 'An economic security council', in: H. W. Singer and R. Jolly (eds), *IDS Bulletin*, 26 (4): 20–27.

Marshall, T. H. (1950) *Citizenship and Social Class.* Cambridge: Cambridge University Press.

Oliver, D. and D. Heater (1994) *The Foundations of Citizenship.* Hemel Hempstead: Harvester Wheatsheaf.

Opeskin, B. R. (1996) 'The moral foundations of foreign aid', *World Development*, 24 (1): 21–44.

Rawls, J. (1972) *A Theory of Justice.* Oxford: Clarendon Press.

Riley, D. (1992) 'Citizenship and the welfare state', in: J. Allen, P. Braham and P. Lewis (eds), *Political and Economic Forms of Modernity*, pp. 179–228. Cambridge: Polity Press for The Open University.

UNRISD (1995) *States of Disarray: The Social Effects of Globalization.* Geneva: United Nations Research Institute for Sustainable Development.

World Bank (1995) *Global Economic Prospects and the Developing Countries.* Washington, DC: World Bank.

8 Partnerships, Inclusiveness and Aid Effectiveness in Africa

B. J. Ndulu

Introduction: Aid and African growth

The international development community and the intended aid beneficiaries in recipient countries have expressed grave concern about the ineffectiveness of aid in poor countries. These countries have failed to achieve sustained growth and poverty reduction in spite of protracted and high levels of foreign assistance. The bulk of these countries are in Africa. Five main symptoms of development assistance ineffectiveness are often referred to in the discourse on this issue.

First is the failure of aid to induce sustained growth in the past three decades despite large inflows of external assistance (Mosley, Hudson and Horrell 1987). Figures 1 and 2 tell the main sad story of an apparent failure of a sharp rise in aid during the 1970s and 1980s to spur sustained growth, which is one of the key measures of aid effectiveness. This failure of growth, in turn, was behind the very slow progress in poverty reduction in the region, another key measure of aid effectiveness.

A persistent, high level of growth is essential for starting up the motor for self-sustenance and for reducing poverty effectively. A vast literature now exists on the centrality of growth to poverty reduction, particularly where poverty is broadly shared (Bruno, Ravallion and Squire 1996; Ravallion and Chen 1997). Where poverty is widely prevalent, the most potent strategy for tackling it would be to increase opportunities for the poor and enhance their capabilities to earn a decent income.

Using data from Collins and Bosworth (1996), Ndulu and O'Connell (1999) decompose the sources of GDP growth for a sample of 21 sub-Saharan African economies, accounting for nearly 80% of sub-Saharan Africa's GDP and population, into capital per worker and productivity residual during 1960–94. They then compare this decomposition to 45 other developing countries and to 22 industrial countries in the data set. Three characteristic features of growth emerge for the sub-Saharan Africa economies. First, Africa's long-term growth has been slow relative to other developing countries, with a sample difference in average growth rates for the 34 years of 1.7%. In

V. FitzGerald (ed.), Social Institutions and Economic Development, 143–168.

per capita terms the difference is wider at nearly 2.2%, on account of Africa's higher population growth. Second, from the decomposition of the sources of growth, slightly less than half of the growth difference with the other developing countries is due to slower accumulation of physical and human capital; and slightly more than half is accounted for by the slower growth in the productivity residual (Table 1). In this sense a much more prominent role in the explanation is attributed to the productivity residual. Third, during the period of persistent stagnation the contribution of physical accumulation to growth goes down to nearly zero, highlighting the productivity crisis.

Figure 1 presents the time trends of growth, investment rate and investment productivity (the inverse of ICOR). Most striking is the very steep decline of the five-year moving average of investment productivity between 1973 and 1986 from a peak of nearly 28% to about 4%. Investment productivity in Africa for the entire period is only half of that obtained in other developing regions. The deceleration of growth follows a nearly perfect matching trend over the same period, with the growth rate declining from an average of 5% to about 1%. In contrast, the investment rate shows a steep rise in the first seven years of this same period and in fact recorded the highest average for the region over the entire period. This was also the period of a sharp rise in aid, provided to fill the financing gap of investment programmes.[1] The sharp decline in the rate of return to investment implies a very marginal con-

Table 1 Contribution of Factor Productivity to Growth

Region and Sub-Period	Contribution of:	
	Growth in Real GDP per Worker (unweighted average)	Factor Productivity (unweighted average)
Sub-Saharan Africa (21 countries)		
1960–94	0.39	–0.44
1960–73	1.76	0.53
1973–94	–0.44	–1.02
1960–84	–0.41	–1.35
1984–94	–0.43	–0.62
Other developing countries (45 countries)		
1960–94	3.14	0.46
1960–73	2.07	1.34
1973–94	1.65	–0.07
1973–84	1.42	
1984–94	1.18	0.19
Industrial countries (22 countries)		
1960–94	2.68	1.01
1960–73	4.30	1.95
1973–94	1.70	0.43
1960–84	1.72	0.13
1984–94	1.69	0.76

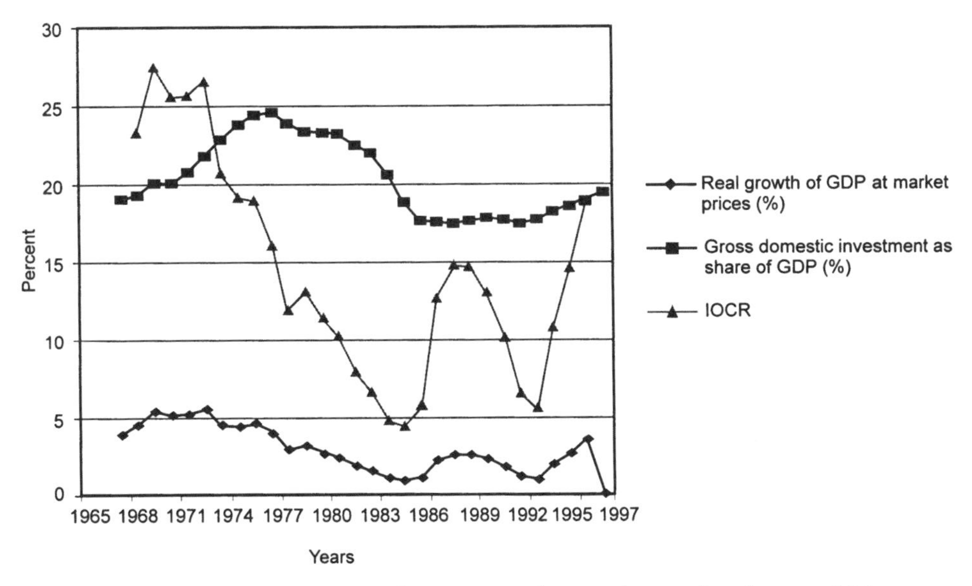

Figure 1 Investment rate, productivity and growth in sub-Saharan Africa

tribution of capital accumulation to growth. Furthermore, to the extent that investment was financed by borrowing externally, it also ushered in a serious debt problem.

We identify three factors as strong candidates among others in explaining the tragedy of low investment productivity in Africa. One is a lack of complementary human skills needed to gainfully use more complex capital. The deepening of import substitution during the second half of the 1970s entailed investment in more complex processes requiring higher skills. This investment was typically pursued in advance of building the requisite human competence. A second likely reason is the poor quality of investment choices and very low utilization of installed capacity. Part of the evidence here is the long death trail of 'white elephants', most of which either lay incomplete or were never made operational. Since aid was then provided mainly for capital expenditures, utilization of the rapidly expanding installed capacity was disproportionately resource constrained. Finally, the steep decline in the productivity of investment could be explained by the prevalence of distortions in policies and markets. Devarajan et al. (1999: 12) argue that poor incentives created by foreign exchange market distortions and high budget deficits could explain why investment is not productive in Africa.

The second concern with aid effectiveness is that aid has tended to displace domestic savings and tax effort, creating impetus for an indefinite reliance on official development assistance. This is the issue of the 'corrosive' effects of aid, which perpetuates dependence or fails to generate a pump-

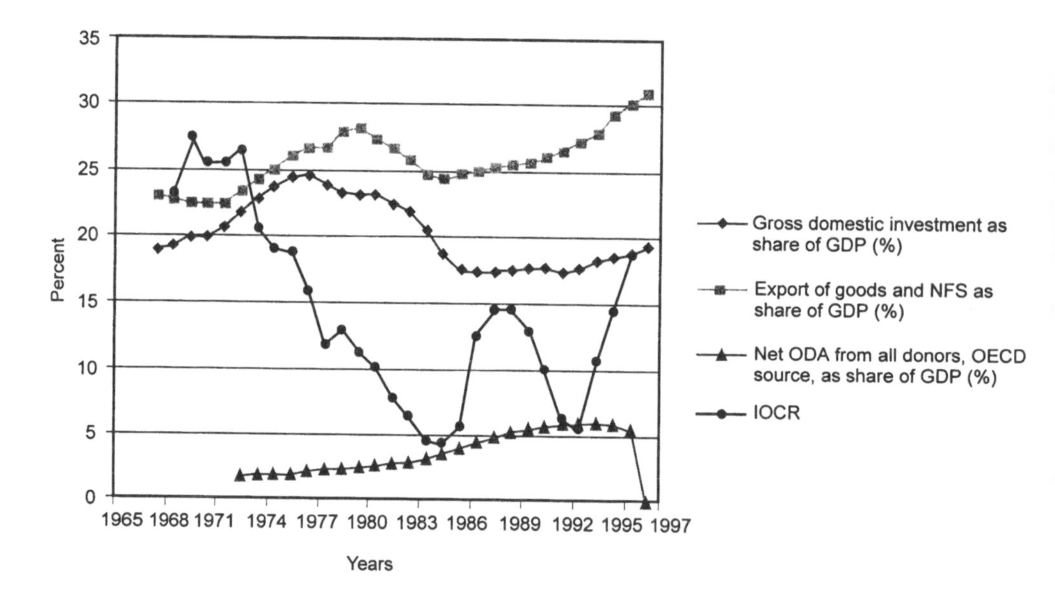

Figure 2 Investment productivity, exports and net ODA

priming effect (White and Luttik 1994, Riddell 1996, Ndulu 1995). This concern builds on a vast previous literature on the negative fiscal response to aid (Griffin 1970, Heller 1975, Mosley 1980, 1987). More recent empirical evidence, however, shows that the displacement is not total, since the relationship between foreign and public savings, though negative, has a coefficient greater than –1 (Hansen and Tarp 2000), and aid has led to increased overall government spending in Africa (Devarajan et al. 1999).

Third, the high intensity and diversity of aid have tended to distort the incentive structure of aid intermediaries and impose high transaction costs in managing aid relationships. Intense aid dialogue and uncoordinated but wide-ranging interventions have imposed high costs for maintaining normal donor relations. Much national energy and political capital is applied to maintaining dialogues on aid and managing a diverse project portfolio. The real opportunity cost of such excessive engagement have been shown to be very large (see Wuyts 1996 for the case of Mozambique and Killick 1995 more generally). Often these engagements compete for attention with domestic development debate and consensus building. In the latter case, high aid intensity has tended to insulate governments, the aid intermediaries, from being accountable to domestic political constituencies (Braetigum 1999) as they focus on dialogues with donors.

Fourth, even when assistance is predicated on fulfilling conditions linked to improvements in the policy environment, it has failed to make any significant difference in the adoption of desired policies, mainly due to donors' failure to enforce conditionality. Lack of such enforcement in turn has resulted from a number of the pressures, such as political clientelism and the desire to avoid the economic meltdowns that could be caused by an abrupt cessation of aid, particularly where aid forms a substantial part of operative and development finance. Further, there is the Samaritan's Dilemma, which donors face when their suspension of aid because agreed amounts have not been spent on the poor would lead only to a worsening of the situation of the poor. Keeping aid flows on a normal footing also preserves the aid agencies' reputation in the face of their political masters (Kanbur 1999).

Finally excessive reliance on external expertise in the form of technical assistance has translated itself into permanent infancy of local capacity to design and implement development policies and programmes. Consequently, developing countries could not exploit the strong virtue that arises from commitment to reforms associated with home-grown initiatives.

The remainder of this paper tries to explain the reasons behind the poor record of aid effectiveness in Africa with a particular focus on the aid relationship as the key influence. The main hypothesis is that aid can be more effective if there is greater inclusiveness in the design and execution of aid programmes and a strengthened citizen's voice as a mechanism to enforce accountability and commitment to results. The next section reviews wisdom garnered from cross-country experience on what matters most for aid effectiveness. Thereafter, the paper discusses a conceptual framework for an effective aid relationship, based on effective accountability systems for achieving desired results and enforcing broad-based preferences. A subsequent section discusses the elements of a partnership approach to aid relationships to enhance aid effectiveness and finally, the paper concludes with suggestions for operationalizing a partnership approach.

Aid effectiveness: What we have learnt from experience

There is now fairly strong evidence that in the presence of the right conditions aid can foster sustained improvement in living standards and effectively reduce poverty. The evidence is based on cross-country experience spanning both successful and less successful cases of development assistance. It also tells us what is crucial for success. The most cited recent work in the tradition of learning from cross-country experience is that by Burnside and Dollar (1997), which closely follows on the vast earlier literature on the same subject but focuses much more on explaining failure. Other prominent contributions are Boone (1996) and Easterly (1997). Three key conclusions from these studies merit attention.

One conclusion is that where good policies exist aid can be effective, leading to an estimated 25% return in terms of growth (Burnside and Dollar 1997). This is a very high rate of return indeed to committed resources. Furthermore, under conditions of good policies, the effectiveness of the overall resources applied also improves, thus raising the stake for putting in place and sustaining a good policy environment. Such an environment ensures effectiveness not only of ODA but also of own resources. Aid applied in a good policy environment is further associated with sharp declines in infant mortality, a proxy for social well-being, while it is not in a poor policy environment. A key implication flowing from this conclusion is that selective allocation of aid to countries that are both poor and have a good policy environment would result in the highest impact of aid on poverty reduction worldwide.

Second, and on the other hand, these studies found little association let alone causation between the application of ex ante donor conditionality and improvements in policy (Burnside and Dollar 1997, Lesink and White 1997, Collier 1996). However, where a domestic constituency for reform exists, adjustment lending did emerge as a useful mechanism to catalyze the reform process. It helps reformist governments stay the course sufficiently long to achieve positive results. These results in turn help to lock in a virtuous circle of reforms backed by a strong constituency of beneficiaries. The more inclusive the range of beneficiaries, the higher the probability of a sustained reform process. This observation is strengthened by the evidence at the micro level from studies showing that when projects are conceived, designed and executed with the participation of local populations, they tend to be more successful than those without such participation. This finding underscores the critical importance of 'ownership' (Isham, Narayan and Pritchett 1995). Results of Isham, Kaufmann and Pritchett (1997) again confirm the critical importance of a strong citizen's voice in ensuring better project performance through participation and better governance.

The microeconomic evidence of better performance where inclusiveness is applied is corroborated by similar evidence at the macro level. Following the categorization of Bratton and Van de Walle (1997), Ndulu and O'Connell (1999) disaggregate growth performance by regime distinguishing among multi-party systems, single-party systems (whether competitive or plebiscitory) and military oligarchies. The classification applies to 1988; with relatively few exceptions it was established by the mid-1970s. At this level of aggregation, for the first generation of independence the multi-party systems were richer on average in 1960 and at every subsequent point. Equally striking is their contrarian growth experience: on average, the multi-party systems diverged from the rest, avoiding the extended contraction that was so prominent a feature of Africa's overall experience. Within the authoritarian group, the countries that were military oligarchies in 1988 were poorer on average at the outset and they remained so throughout; neither of the authoritarian subgroups avoided the dramatic post-1970s slowdown of growth.

Finally, where growth and poverty reduction were adopted as key objectives and strategic interests de-emphasized, returns to ODA tended to be higher. Undue emphasis on commercial and geopolitical interests, profligacy in the use of aid funds, and aid to corrupt or uncommitted governments for strategic reasons were singled out as a major contamination to the developmental objectives of aid programmes (Lesink and White 1997). The end of the Cold War augurs well for refocusing aid on supporting sustained growth and poverty reduction.

There are important outstanding issues however, which the international development community must resolve as the new approaches implied by the foregoing are implemented. Hansen and Tarp (2000) undertake a detailed critical review of the Burnside and Dollar analysis, posing important questions on the implications of their conclusions for development cooperation practice. One issue is that of selectivity. While the Burnside and Dollar study suggests that programme and project support should concentrate in countries committed to sustaining an improved policy environment for maximum effectiveness – a trend that the SPA evaluation report (OED 1998) shows already to be in effect – there is a question on what to do with those countries without such an environment and where poverty abounds. Some suggest that such countries be assisted in creating a conducive policy environment, though not through conditionality as it has proven ineffective in initiating reforms, but by supporting emerging local constituencies for reform. The main form of assistance would be technical assistance aimed at resolving issues requiring information and knowledge, in addition to humanitarian assistance. Yet this approach would require better foresight and monitoring of developments in these countries.

Furthermore, thresholds for good policy environment will need to be agreed upon and country monitoring systems of high integrity and credibility set up. Controversies still abound as to what is an appropriate boundary for a good policy environment to serve as a benchmark for evaluating individual countries. There are systems for such assessments at present but there is no definitive consensus yet on a single definition of its elements and their weights in the total index. Early efforts are needed to arrive at a consensus of what this composite definition of a good policy environment should be.

Effectiveness of government as aid intermediary: A conceptual framework

Principal-agent analysis

Ravi Kanbur (1999) succinctly summarizes the aid relationship as a principal-agent problem. The theory of donor-recipient relationships, as developed in the economic literature derives from the principal-agent analysis. The standard way in which the relationship is modelled is in terms of a Stackleberg leader-follower interaction. The donor is the leader and decides on the level and composition of aid. The recipient is the follower who, taking as given the level of aid, decides on actions (for example, public expenditure patterns or trade policy) which affect outcomes for the recipient (access of the poor to education, economic growth). But the donor also values these outcomes and chooses the level of aid to influence the choice of actions by the recipient and hence the outcomes for the recipient. The level of aid is thus chosen to maximize the donor's preferences, subject to the reaction function of the recipient, which in turn comes out of the recipient's preferences and shows the actions the latter would choose for each level of aid (Kanbur 1999). Differences of preferences/priorities in development or imperfect information about the real intentions of the agent presents a basis for the problem of time inconsistency in the aid relationship, particularly where mechanisms for enforcing the principal's preferences are ineffective. Since in the typical case aid is additional to domestic resources, fungibility of aid adds to the likelihood of the government's composition of expenditure diverging from that preferred by donors.

The simplest versions of the theory assume the donor and the intermediary to be unitary entities, represented only by a set of preferences. They rarely incorporate the processes for determining the preferences and accountability arrangements, much as these influence the enforceability of contracts entered between the principal and the agent. In most models it is assumed that the donor is more concerned with the poor (the ultimate beneficiary) than the intermediary is. In general, all that is needed to make contracts between the principal and the agent tenuous is that their preferences are different. The donor typically relies on aid conditionality to enforce compliance – by pricing out aid in terms of the actions taken. The recipient may prefer unconditional aid but there is no choice (Kanbur 1999).

Pedersen (1998) carries out a detailed conceptual analysis (through simulations) of a variety of strategic aid interactions in the principal-agent framework. Pedersen groups them as a *passive leader donor* (donor makes decision and lets agent adjust in the political context; untied aid); the *active leader donor* (applying conditions in attempt to enforce the donor's preferences); and the *follower donor* (where the donor makes its decisions after the agent has). All these varieties face the same problem of time inconsistency because of either the agent's hidden information or hidden action.

We employ here a slight variant of the traditional application of the principal-agent problem to aid relationships. In this variant, we consider government as an aid intermediary (agent) with preferences that do not necessarily reflect those deriving from an inclusive domestic political process for determining development priorities. The government intermediates between the principal (the financier) and the electorate who are the ultimate beneficiaries of aid and indeed the main contributors to the total public resource envelope. In contrast to a government in an authoritarian regime, in a democratic system the government, in effect, is accountable to two 'principals' for outcomes from applying the resources availed through aid and the country's own resources. The electorate is able to vote out a government in case of failure, while the donor retains the power to withdraw aid if the intermediary does not deliver on promises. Where there is coincidence in the preferences of these two principals, the leverage for enforcing compliance is greatly enhanced.

This variant is similar to another part of the theoretical literature, in which the recipient is modelled not as a unitary government, but as a combination of interest groups interacting amongst themselves and the government represents a democratically-processed set of preferences. The donor, through aid flows, then tries to influence this process of domestic political economy, strengthening the hand of one group against another, ensuring that some actions are more likely to be taken and hence some outcomes are more likely. Adam and O'Connell (1998) and Coate and Morris (1996) are good recent examples of such work. The problem is both the limited availability of channels for exerting such influence and the sustainability of the strategy as the fortunes of interest groups change.

Conditionality is a 'commitment technology' that overcomes the time inconsistency inherent in these problems, but only if it is strictly adhered to. If there is no means for ensuring enforcement of conditionality, and hence a credible threat in the event of violation, the outcomes would essentially depend on the behaviour of the agent. Donors are typically unable to enforce conditionality on governments except where the conditions coincide with the concerns of the beneficiaries, as pointed out earlier. In order to ensure effectiveness of conditionality one needs to minimize the divergence in preferences and institute a credible system of accountability and threats. In a democratic system, as stated earlier, the government as an aid intermediary reports to two principals, financiers and electorates, and hence operates in a system of dual accountability. This is in contrast to the situation of an autocracy, where a single principal exists, the other having no voice. In a democracy if there is a coincidence of preferences between the two principals, the dual accountability system can be collapsed into one.

Several mechanisms exist for locking in the positive changes achieved through reforms. External agencies of restraint have received most attention to date (e.g. policy conditionality and reciprocal threats). However, a number of potentially potent domestic restraints can also be used. For example, many countries have bestowed statutory autonomy to central banks to enable non-interference in the pursuit of prudent measures for macroeconomic stability. Such measures include those to ensure price stability, preserve the value of the local currency and enforce sensible behaviour among financial institutions so as to ensure the financial sector's stable and efficient operation. The central bank can also play a supportive role with the treasury to enforce fiscal discipline.

Another example of an effective approach to locking in positive policy changes is to empower the beneficiaries and the press to play an active role as watchdogs against reversals or erosion of benefits. Governments could institute fora for open dialogue with farmers' associations, exporters' associations, chambers of commerce and other business associations. These and other means of exploiting the growing freedom of press and other accountability instruments under the more open political systems can be used effectively to discourage pressures from vested interests for unwarranted reversals of good policies.

Behaviour of the aid intermediary under a weak authoritarian regime

African governments initially acquired power largely through bargains with external rather than internal actors. The interventionist stance adopted by most African governments at independence had its logic in the development paradigm of the day, which emphasized addressing the preponderance of market failures and the nascence/fragility of institutional structures which hampered development. The dominant peasant economy was considered not only technologically backward but also as lacking the requisite dynamism for autonomous development. It was argued that the state needed to play a central role as the principal agent for modernizing the economy (Ndulu 1986). Government was therefore to use its fiscal powers, the external resources channelled through it and indirect controls on private-sector resource allocation to this end. The government promised development in exchange for the right of the state to maintain a centralized authoritarian system of governance. Thus, a social contract was struck early in the post-independence period that traded the right to open governance structures for patronage and a promise of rapid growth in what has been referred to as 'developmental autocracy' (Gordon 1990).

Under authoritarian rule, development policy has tended to be captured by a narrow group of elites operating under relatively weak institutional constraints. The political legitimation of this paradigm drew on the liberal assumption that the state is a neutral and even a benevolent arbiter among the different interest groups to further national interest in economic growth, efficiency and social welfare (Sandbrook and Barker 1985). Donor support of all types was channelled through the state with the same understanding.

The private sector in these economies in essence was part of the covenant with the state that granted it protection, access to subsidized resources and significant rents derived from protection. In many cases, a strong symmetry of interests existed between the public sector and the politically-connected private sector as far as price distortions, protective measures and access to subsidized resources were concerned. Economic nationalism was promoted through both state enterprise and the encouragement of the indigenous private sector through the above-mentioned preferential arrangements.

Applying a variant a variant of Olson's (1982) 'encompassing interests' paradigm to fiscal policy choices under authoritarian rule, the main explanation for poor growth performance in authoritarian regimes offered by Ndulu and O'Connell (1999) is that the ruling elite in these regimes sacrifice the general interest in order to extract rents and retain power. Outright predation (à la Mobutu) can emerge when rulers believe that wealth accumulation outside the control of the elite will increase political contestability. This contrasts sharply with high-growth cases such as Botswana and Mauritius, where national-level policy has been contested through regular elections and meaningful legislative roles, both enforcing a discipline of accountability that benefits the electorate more broadly. A narrow elite over-taxes and under-provides public goods, sacrificing overall growth for the sake of transfers to itself; broad participation internalizes the general interest in growth. This analysis is readily adapted to motivate policy patterns widely observed in African countries.

The compass of the government is the exogenous fraction, f, of national income directly controlled by the elite. Current consumption of the elite is given by $C = fY - I$, where Y is the (exogenous) current national income and I is infrastructure spending. In the absence of aid the choices open to government are constrained by domestic revenue. The opportunity cost of a unit of infrastructure spending is therefore a unit of current consumption. Future consumption of the elite, $C_F = fY_F(I)$, increases with infrastructure spending since today's road-building increases the economy's future productive capacity.[2] Leaders of autocratic regimes with high discount rates on account of the high risk perception of their continued rule will maximize transfers to the narrow elite at the expense of investment. A more encompassing regime would choose to expand future productive capacity.

The basic model as applied in Ndulu and O'Connell (1999) poses a policy choice in the allocation of budgetary resources between transfers to a narrow elite in a patronage system and developmental spending represented here by expenditure on infrastructure. Tax revenues are thus split between public infrastructure spending and per capita transfers of size T to a favoured group of size f (with the population normalized at 1).

$$fT + \text{infrastructure spending} = tB(t), \qquad B'(t) < 0 \qquad (1)$$

Domestic taxation is distortionary, so that a rise in t shrinks the tax base $B(t)$ and creates a deadweight loss. For fixed t, growth effects emerge over time through the undermining of investment in the taxed activity.

To motivate the choice of t and T we think of the government as maximizing the welfare of the favoured group. As long as infrastructure spending is positive, some amount of distortionary taxation cannot be avoided. In this situation a tiny transfer creates a finite additional distortion on the margin and is therefore socially inefficient. It follows that a government with fully encompassing preferences – in fact, even one with 'sufficiently' but not fully encompassing preferences (see Boone 1996) – will forego transfers. There is some level of f, however, below which the concentration of economy-wide revenues into the hands of the favoured group will justify transfers. Narrower power structures will generate larger transfers and slower growth.

The model accommodates broad features of African economic policy – in particular, heavy discrimination against international trade and widespread use of quantitative controls in preference to price interventions. Given the exclusion of peasant interests from the favoured group and the low administrative cost of taxing external trade, transfers from export agriculture are often large.

We extend the basic model (equation 1) below to include the line items for budgetary financing. The extended budget constraint therefore includes net inflows of foreign aid and borrowing from the banking system or inflation tax revenue. Given the fairly undeveloped market for government securities in the region, we exclude non-bank borrowing from the significant sources of budget financing.

$$fT + \text{infrasructure spending} = tB(t) + FA + p\text{m}(p)$$
$$p\text{m}'(p) > \text{o for } p < p^* \text{ and } < \text{o for } p > p^* \qquad (2)$$

where p = inflation rate, m = real money demand, p^* = optimal inflation rate beyond which revenue from inflation tax declines in absolute terms.

In this extended version of the basic model, aid and revenue from inflation tax provide additional resources to the government to finance infrastructure and transfers to the favoured groups. Until the early 1990s foreign assistance was readily available and was channelled through the state apparatus in exchange for what leadership elites could supply. This did not include rapid development, which occurred in some cases but appears to have been incidental to aid flows (Burnside and Dollar 1997). In equation (2) untied or fungible aid inflows can enhance growth by allowing a reduction in distortionary taxation or financing public infrastructure. For a sufficiently encompassing government this is what the model predicts (Boone 1996). It is easy to see that if conditionality was effective a similar effect would emerge if aid financed roads and agricultural research.

Yet a consensus is emerging in the empirical and case-study literature (e.g. World Bank 1998) that aid has had little effect on policy outcomes in Africa and that its contribution to African economic growth on the margin has been minimal or even negative. A central reason for this may be the lack of feedback from development outcomes to aid flows. A rise in unconditional aid should increase transfers where they are already active. Moreover, unless the government is fully encompassing, there is some level of aid above which transfers will be initiated in preference to further tax cuts or increases in public infrastructure spending (Adam and O'Connell 1998). Aid does not directly reduce growth in these cases, but policy distortions will appear, ex post, to be robust to inflows. In the absence of effects on succession, slow growth is a locally stable political equilibrium (Collier and Gunning 1997). The primary effects of aid in this situation may operate through domestic competition for politically-motivated transfers and the power to dispense them.

From the perspective of governance, seignorage revenue plays a similar role to that of unconditional or fungible aid. As long as it stays on the correct side of the inflation tax Laffer curve, an all-encompassing government can use such revenue to reduce distortionary taxation or raise spending on public infrastructure. However, beyond the optimal inflation tax rate the government moves to the wrong side of the Laffer curve and revenue from this source falls in absolute terms. Furthermore, tension between the government and the favoured groups may ensue from the effects of high inflation as the latter experience real erosion of wages or profits; and distortions in the incentive structure lead to reduction of the tax base, particularly as the local currency appreciates and exports decline.

During the first half of the 1980s many African countries outside of the CFA zone resorted to this kind of financing, the majority of them overshooting the optimal inflation tax rate. They took this course of action partly to cushion themselves against the consequences of fiscal crises and a temporary decline in external aid inflows associated with the global recession and the debt crisis in the first half of the eighties. In a number of cases, however, excessive use of inflation tax persisted well into the 1990s. Adam, Ndulu and

Sowa (1996) provide examples of such overshooting for Ghana, Kenya and Tanzania. Absence of a strong and independent central bank was typically a reason for lack of restraint to stay on the correct side of the curve.

Unsustainable development autocracy and the seed for a more inclusive governance

Within the logic of the weak autocracy framework of governance, there are two main underlying trends which make the demise of the regime imminent. First is the time inconsistency of the system of rent extraction in the absence of growth due to distortionary effects of taxation and the fact that very soon these types of distortion get the system to the wrong side of the Laffer curve. Second there is an assumption that the multitude of the unorganized interest groups, including the peasantry is politically passive and acquiescent to the benevolence of the state, even if state actions are geared towards catering for the interests of favoured and vocal groups. The regimes in power totally underestimate the potential of grassroots responses to frustrate predation.

The revenue base is the most important political resource for servicing patron-client networks. Its erosion through faltering revenue collection undermines the capacity of the political patrons to service their networks of clients. Economic stagnation due to under-investment in productive capacity and the increased proportion of the economy that goes underground negatively impacts the tax base. The steady preponderance of controls to create rents for the favoured groups is the main reason behind the mushrooming of parallel markets and illegal cross-border trade to evade them. This development not only undermines the fiscal base of the state but also significantly reduces rental incomes which hitherto were available to service patron-client networks.

The other undercurrent is the latent tension between grassroots interests and the purported national interests advanced through state policies. African social structures are characterized by the dominance of relatively autonomous networks bound by kinship, tribe, religion, race or community ties. These networks span across rural-urban boundaries in what Hyden (1986) calls the 'economy of affection'. The state is structurally superfluous and it is only acknowledged to the extent that its actions are considered beneficial to the interests of these networks. Where government policy is deemed unbeneficial or a threat to their interests, these networks through their own communication systems have frustrated such policies through use of a variety of 'exit options'. These include participation in parallel markets, illegal border trade and even informal banking systems. These grassroots networks have proved to be a potent force of frustration of predation, when their own interests are ignored.

The roots of recent changes towards a more encompassing governance system lie in the above developments. Conditionality associated with adjustment finance has to a large extent amplified these latent pressures for change and catalyzed the process. By the time the externally enforced adjustment had taken effect, the traditional role of the state had become unsustainable. The government had overextended itself relative to its financial and managerial resources. Society's cynicism of the effectiveness of development institutions had set in. Pressure for dismantling controls was being expressed by rampant evasion through vigorous parallel markets, which in most cases were subsequently formalized and made more efficient under reform programmes. Downsizing of government operations was essentially a forgone conclusion given the emerging unsustainable resource gaps and the rising debt service burden in the *fiscus*. Meanwhile, the global change of view towards a minimalist government gathered momentum and was brought home through aid relationships and international governance. Furthermore as democracy has set in, the threats of removal of a regime from power rather than the use of latent exit options became more dominant.

Moving more positively into the twenty-first century through a tripartite partnership

The changing governance context of development assistance in Africa

High aid intensity has in the past tended to generate perverse incentives for accountable behaviour of governments. By providing additional financing room for sustaining patron-client networks, it has tended to slacken pressure on governments for adopting inclusiveness in the political process. Aid has also reinforced dependency psychology for development management by undermining the emergence of local competence for such management. The main culprit in this latter problem is the intensive use of technical assistance instead of strengthening local capacity. Enclave project management systems have often been set up to deal with existing capacity weaknesses; and in some cases (Tanzania and Uganda) up to 70% of aid has been given outside of budgets through parallel aid management systems.

A rapid process of liberalizing political systems is under way in Africa, with greater acceptance of democracy and increased pressure for devolution of authority to sub-national governance entities. This adds pressure for greater inclusiveness across stakeholders in designing development programmes and for fundamental changes to be made in accountability systems to make them fit with more open governance structures. This development poses perhaps the greatest challenge to reformulating the aid relationship. The current dual accountability system of recipient governments to donors and to the local constituency may need to be reconfigured into a single integrated system. The growing emphasis on a partnership approach and local ownership is one innovative response. But to make partnership and local ownership a reality, the immense challenges of surmounting attitudes and changing procedures for aid management first have to be faced.

Accountability systems have been fragmented, revolving around a dialogue between the government and the multitude of donors typically involved in a specific African country. The emphasis has been on external restraint to imprudent behaviour by the main intermediary of aid – the government – and seldom on internal mechanisms for restraint. This arrangement was consistent with the prevalence of autocracy, but is now facing serious challenges as open polities and inclusiveness in the system of governance are being widely adopted and donors and domestic stakeholders are becoming increasingly intolerant towards corruption. I already alluded to the problem of dual accountability, as well as the need to resolve it through an integrated approach. Much more sanguine conclusions regarding the ineffectiveness of external conditionality in begetting change and reform buttress the shift towards greater reliance on domestic constituencies to motivate and sustain change. Increased attention is also warranted to promote the role of citizens' voice in begetting accountable behaviour among those charged with managing the development process.

Developing countries are switching from a directed approach of resource allocation to a more market-oriented one. The intention is to make resources available to the most productive users and to support a growing contribution by the private sector, including enterprise and local community initiatives. This change implies a need to allocate aid following established incentive structures to complement the strategic impacts mentioned earlier. However, it questions the continuation of using the government as the sole intermediary of aid and calls for diversification of channels of support. Yet in this respect there is a need to harmonize procedures and provide information as to appropriate donor entry points in order to preserve the coordination and policy roles of public authorities. This can be done without prejudicing the strong advantage of utilizing more decentralized channels of assistance, including non-governmental organizations and domestic financial institutions.

Imperatives for more effective development assistance

The foregoing allows us to identify five critical areas for improving the effectiveness of development assistance. First is to encourage and support the development and sustenance of an environment that is conducive to raising the returns to resources applied for development. It entails inter alia putting in place good policies, maintaining a stable civil environment, improving supportive infrastructure and strengthening the capacity of the state to judiciously and prudently conceive and execute development programmes. These considerations are in line with a focus on assisting governments in providing public goods to enhance the overall efficiency of resource use and to improve conditions for attracting private capital. In this, and given the fungibility of development finance, it is wise to focus on the whole public expenditure programme to achieve effectiveness. Developing a rational expenditure plan, which can partly be financed from aid, permits realization of overall effectiveness.

Second is to redefine aid relationships in a manner that permits recipient countries to play a lead role in designing, implementing and monitoring development programmes, in cooperation with donors. Local ownership, which has been shown to provide better results when encouraged, is at the heart of the new relationship. It is important to emphasize here that local ownership should be inclusive of civil society and not just government. The strength of this type of relationship lies in its ability to elicit commitment and create conditions for local pressure to ensure accountability and restraint from wasteful use of resources. For this to occur however, three important preconditions must be fulfilled. One is adequate capacity and resolve on the part of the recipient country to take up the lead role. Two is the development of an open political system and upgrading of civil liberties to bring the policy process into the public domain and subject it to citizens' voice. Three is donors' willingness to wait for and adhere to coherent strategies on which they should base their specific assistance. In this last respect, promoting local ownership is consistent with better aid coordination and will help avoid the problems currently associated with proliferation of enclave projects and unmanageable recurrent costs profiles resulting from this proliferation.

The third critical area concerns enhancing complementarity between public provision of services and participation of the private sector and local communities in the provision of social services. This concern arises partly from recognizing the limitations of the resource envelope of the public sector and partly from the enhanced effectiveness of service delivery when beneficiaries are able to enforce a higher level of accountability. Greater involvement by local authorities in the provision of social services should improve effectiveness of delivery and enhance user contributions to expenses, thereby relieving the government's budget. Countries can also benefit from higher levels of cost-efficiency in health and education services delivery by the private sector compared to the public sector. A clear division of roles is also emerging between the private and public sector in the provision of social services, with the latter being more dominant in basic services particularly in the rural areas.

The fourth area for improvement is in adapting development cooperation to the changing environment. Foremost among these changes are the increased demand for more effectiveness in utilizing declining aid budgets, the growing dominance of private capital in development finance, the deepening integration of developing countries into the global economy and the continuing global trend towards increased market orientation and more open governance structures. These offer new opportunities as well as challenges to development cooperation. Most important among the required responses is forging stronger complementarity between ODA and the rapidly growing private capital. Infrastructure has been the exclusive domain of government in African countries. But new opportunities for leasing public infrastructure for private operation and arrangements for building, operating and transfer of such infrastructure as roads open up new opportunities for private-sector participation on the continent. The end of the Cold War allows minimization of contamination of aid objectives for political or strategic clientelism.

Finally, the presence of a stable and fair civil environment is crucial to ensure that the destruction of human and physical resources does not continuously set back the clock of development. In this regard the predominant reactive approach to crises has tended to be too late. A more preemptive tactic is needed to avoid wastage. In the minds of investors the legacy arising from civil strife and wars tends to linger on for long periods of time once they have occurred (Collier 1996). Even more disconcerting are the spillover effects of civil strife, which spread beyond the countries concerned. The burdens of multiple simultaneous transitions from civil war to peace are awesome and add to the pains of the other transitions that many of these countries are making, from controlled to market economies and from autocracy to democracy. The virtues of current economic reforms have often been stifled by these difficulties and have tended to prompt investors, local and foreign alike, to adopt a 'wait and see' stance due to uncertainties linked to past bad reputation and risks of asset loss. Although the primary responsibility for avoiding conflict lies in the first place with the concerned countries, the international community can help promote civil liberties, with cultivation of a culture of tolerance as a pre-emptive measure.

Partnership in practice

Raising the effectiveness of official development assistance requires a candid and thorough reappraisal of approaches to the provision and utilization of aid by all parties concerned. Such reappraisal must learn from past experience and take into account the major changes taking place within the recipient countries and the donor community, as well as in the global market. On the one hand, a review is called for of donor practices in designing and implementing programmes of assistance and in creating conditions for efficient and transparent use of committed resources on the side of recipients. Of particular importance now is to create a common framework for designing development programmes that reflect priorities arrived at through inclusive processes and a monitoring/accountability system that emphasizes results and transparency in resource use. The development dialogue needs to bring together donors with government as aid intermediary and civil society in recipient countries in agreeing upon development priorities; and the accountability system should be unified in a similar fashion.

Local ownership and partnership

We propose here some mechanisms for a policy process and forms of aid management that would enhance local ownership of aid programmes. The use of national development programmes, sector development programmes and public expenditure review processes, all led by recipient governments with a broad range of stakeholder inputs, is key to a successful framework of the new partnership. Furthermore, devolution of responsibilities to lower levels of government and to communities for service provision through decentralization deepens the new ownership framework if backed by improved managerial capacity and accountability systems. Ultimately, however, changing attitudes towards the aid relationship rather than the technical difficulties is likely to prove most onerous in this new approach.

Figure 3 presents a prototype of a new partnership framework as it is being developed in Tanzania. This is an ideal framework, which the country ultimately hopes to implement in its entirety. An assistance strategy developed by the government through a consultative process with donors and civil society is the fulcrum for defining development priorities consistent with the country's long-term development vision. The sector development programmes, including investment programmes and their recurrent cost implications, are guided by this strategy. A public service reform programme includes the strategies and action plans for improving the effectiveness of public service delivery, enhancing accountability systems including results orientation and an institutional capacity-strengthening programme for public management. This programme further incorporates improvements in financial management and greater transparency of budgets. More inclusive public expenditure reviews, analytic work and consultative group meetings provide inputs into the process of prioritizing development activities and the development assistance framework.

Having estimated the costs of the development programmes and the projected resource envelope, including planned donor support consistent with priorities in the assistance strategy, a medium-term expenditure framework is developed reflecting the priorities agreed through the inclusive process. Under the public expenditure review process a working group led by government and involving donors and civil society is the main forum for

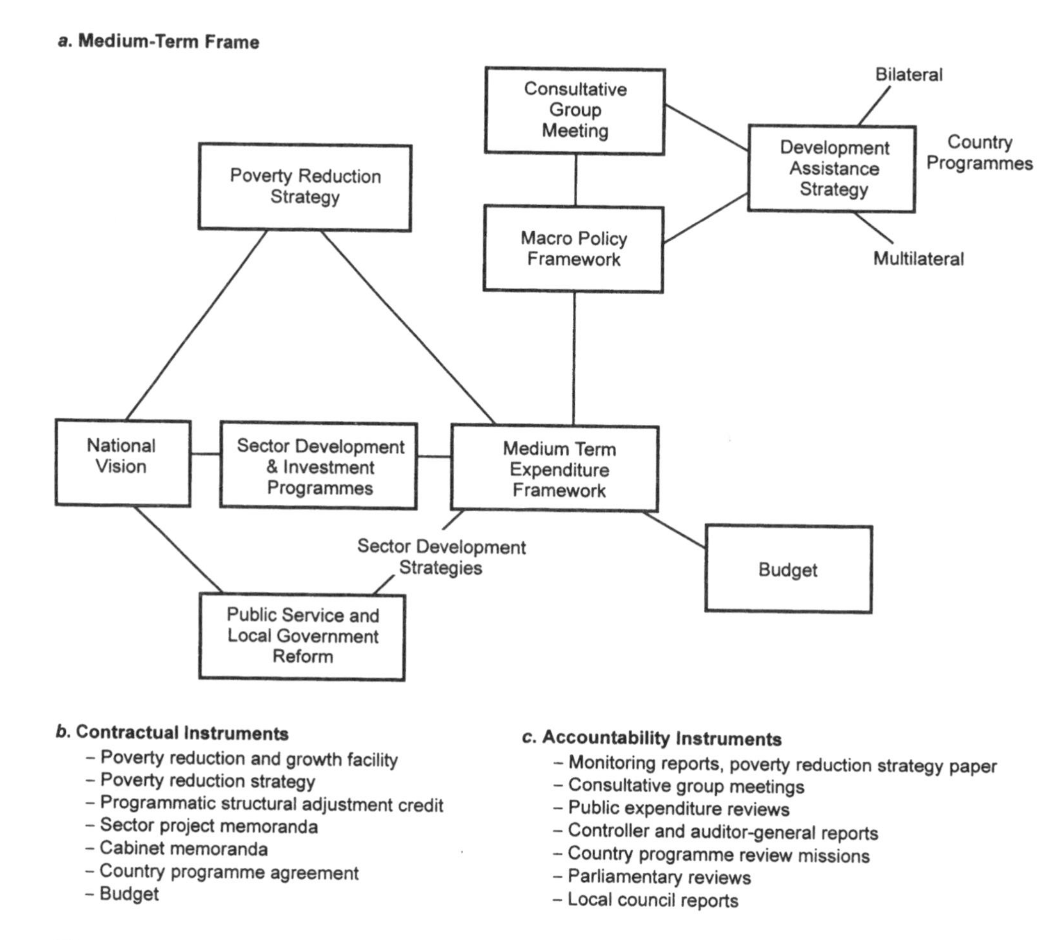

Figure 3 A prototype implementation framework in a partnership approach

agreeing on the expenditure plans, which should be consistent with a desired macroeconomic framework. The individual donor-country programmes are agreed on within the development framework and integrated into the exchequer system. Each year's budget frame is approved by the parliament following this process. Implementation of the overall programme is monitored in relation to the government-led overall programme and the inclusive public expenditure review process. The reports by the controller and auditor-general are the main instruments for ensuring a unified accountability system in the public domain. The bilateral country programme reviews and the multilateral portfolio reviews are carried out based on the findings of this common framework for reporting and evaluation.

Accountability

In an open governance system, the tension between the two channels of accountability, one to donors and the other to electorates, has become more pronounced and needs resolution. A partnership approach where donor preferences are integrated in the processing of priorities prior to national approval of programmes holds good promise. Yet this must be complemented by clear monitoring targets and credible evaluation systems, which will offer the necessary comfort to electorates in both donor and recipient countries. A more credible, accountable system will also encourage integration of donor resources into national programmes derived from a national consensus and promote increased disbursement via budget or programme support. The need for a framework for resolving competing interests through partnership approaches and open discourse cannot be overemphasized.

The main issues here relate to integrating the accountability systems and avoiding operating outside the budget system (for public support). Integration would be helped by adopting a system of national approval of aid programmes, with prior inputs from or in consultation with stakeholders and donors. This system may be administratively cumbersome, however, and time may be needed before parliaments as representative bodies can effectively process inputs independently. The second best process remains that of wide consultation. If inclusiveness is to become a reality, it is imperative that a credible and transparent system of accountability for resources and results be in place to facilitate the reporting requirements of donor agencies to their own governments and electorates.

The required changes in procedures on the donor side are large and the transitional difficulties should not be underestimated. However, unless these changes are undertaken the much talked about 'new partnership' is unlikely to materialize.

Improving efficiency of allocation and utilization of aid

Declines in aid budgets and the clamour for enhanced effectiveness require that efficiency in allocation, disbursement and application assume much more importance than simply as levels of support. This, in turn, calls for a more focused and coordinated approach to providing and using assistance so as to elicit strategic impacts and avoid wasteful duplication. Since strategic guidance for allocation can only occur where *overall* programmes are well designed, allocation of aid ought to take place in the context of such programmes. The role of the medium-term expenditure framework in improving the predictability of budgets and coordination of external assistance is crucial. It should lead to a more cohesive and productive allocation of aid resources, which typically are fungible in the context of overall resource allocation. Furthermore as argued before, this approach requires a fundamental change in the 'primacy' of drawing up and managing development initiatives, from donor-centred to an inclusive process led by recipient countries. Donors can still exercise *selectivity* based on clear signals regarding recipients' ability to take up the lead role.

There is also a need to develop a credible system for monitoring effectiveness of disbursement and utilization of resources availed. A result-oriented approach to such monitoring rests squarely on public expenditure reviews, which should not only seek to establish whether money was used for the intended purpose but also whether the undertaking achieved what it intended to do. Nor should these reviews solely serve the purposes of the donor's side; they ought to be internalized locally as a measure of efficiency and a clear demonstration of value for money to the local development constituency. More broadly however the performance assessment could focus on the key objectives of assistance, which may include growth performance, a set of social well-being indicators, impact on domestic resource mobilization and a well defined assessment of strengthening local human and institutional capacity to carry on in the future. To the extent that simplified and integrated procedures for accounting and reviews can be developed, they would help improve compliance and transparency.

Developing institutional capacity for public service delivery

Given the pivotal role of recipient governments as aid intermediaries, it is fundamental to stress that their capacities to develop, engender wide consensus and to implement desirable development initiatives are crucial for raising effectiveness of development assistance. The focus in the past decade has been on rationalizing the role of governments in the development process by retrenching away from activities that the private sector can do best. Lack of competence and weaknesses in institutional structures remain obstacles to state capacity for a more effective but limited role of unleashing local initiative, ensuring a good policy environment and providing supportive infrastructure for gainful activity (Braetigum 1997). Fundamental to strengthening these competencies is the development of technocratic expertise in programme design, assistance in the development of markets where they are nascent, provision of a regulatory framework to counter failures or incompleteness of markets and positioning of countries to exploit the vast and rapidly developing global knowledge base of development experiences.

Better methods of conveying knowledge will be helpful only if local capacity to absorb and adapt knowledge to local conditions exists on the ground. Equally important is the development of professional ethos in the civil service and appropriate incentive structures for developing and sustaining professional growth. Also critical is a strengthened legal framework to permit expeditious settlement of commercial disputes so as to reduce uncertainties that hinder private sector growth.

The thrust of the challenges here relates to enhancing local ownership of development programmes, not only as a means of reducing reliance on technical assistance but also as a way of improving aid effectiveness. Sustaining reforms and creating conditions for longer term development require commitment and adequate managerial capacity, not only in recipient governments but also in the population at large, which forms a local constituency for sustained development (Helleiner 1993, Ndulu 1994, 1995).

Internally a governance structure that encourages participation, accountability and freedom of action is conducive to promoting local initiative and flexibility in responding to changing conditions. It also fosters institutional innovation, ensures a broad vetting of economic policy and implementation and reduces risks associated with a non-transparent system of governance. Donor participation in this process of determining development priorities and the means to achieve them needs to assume a posture of partnership to all local actors, not simply governments.

Technocratic capabilities and supportive institutional structures need to be enhanced to enable local formulation and implementation of sound macroeconomic and sectoral policies. Emphasis should be on capacity building for policy analysis and design and monitoring of development programmes. Fundamental to this endeavour is support for localizing preparation of policy framework papers and sectoral programmes and coordinating development programmes and performance monitoring.

Capacity-building initiatives are not confined to governmental entities alone. The private sector, particularly the informal sector, needs to acquire the skills to manage enterprises efficiently. Enhanced institutional capacity for providing information on markets and investment opportunities can go a long way to improve economies' overall efficiency. In the same view, to allow local communities to contribute meaningfully to development, there is a need to strengthen local capacities to plan and execute local development projects. Non-governmental agencies and local governments have received support in this regard, but their activities need to be coordinated with national development strategies.

As countries begin opening up the policy process, the range of participants in policy formulation and monitoring of policy outcomes is widening to include academic and research institutes as well as other interest groups. Strengthening these groups' capacities to contribute meaningfully to policy analysis will broaden the ownership base of development programmes and foster accountability by those entrusted with the immediate responsibility for policy formulation and implementation. Technical assistance that seeks to address this challenge will enable the recipient countries to ultimately wean themselves from excessive reliance on external competence for managing their own development affairs.

Notes

1. In fact real ODA (deflated by the DAC deflator) as a proportion of GNP for 33 countries with populations exceeding 800,000 shows a sharp rise during 1975–85, with a peak reached in 1985 (O'Connell and Soludo 2000). The peak in aid intensity measured in nominal dollars was reached in 1990.
2. The government chooses I to maximize the discounted consumption of the elite: Max $W = C + \beta CF = (fY - I) + \beta fYF(I)$, {S}, where $\beta < 1$ is an exogenous discount factor.

   The government's optimal choice equates the marginal cost of providing I to its marginal benefit. Recasting the problem as the equivalent one of maximizing W/f is $1/f$, reflecting the concentration of its financing in the hands of the elite. The marginal benefit, in turn, is just the discounted marginal product of public infrastructure, $\beta \cdot dY_F/dI$. With diminishing marginal returns to infrastructure this is declining in I. The socially efficient level of infrastructure is I^*, where the full *social* marginal cost (= 1) equals the social marginal benefit ($= \beta \cdot dY_F/dI$). The government chooses instead to trade off infrastructure for higher current consumption ($I^E < I^*$). A more encompassing elite chooses better policy.

References

Adam, C. and S. A. O'Connell (1998) Aid, taxation, and development: Analytical perspectives on aid effectiveness in sub-Saharan Africa. WPS/97-5, January. Oxford: Centre for Study of African Economies, Oxford University.

Adam C., B. Ndulu and N. Sowa (1996) 'Liberalization and seignorage revenue in Kenya, Ghana, and Tanzania', *Journal of Development Studies*, 32 (4): 523–531.

Boone, P. (1996) 'Politics and the effectiveness of foreign aid', *European Economic Review*, 40 (2): 289–329.

Braetigum, D. (1997) 'State capacity and effective governance', in: B. Ndulu and N. van de Walle (eds) *Agenda for Africa's Economic Renewal*. Oxford and New Brunswick: Overseas Development Council, Transaction Publishers.

Braetigum, D. (2000) *Aid Dependence and Governance*. Stockholm: Expert Group on Development Issues (EGDI), Ministry of Foreign Affairs, Sweden.

Bratton, M. and N. van de Walle (1997) *Democratic Experiments in Africa*. Cambridge: Cambridge University Press.

Bruno, M., M. Ravallion and L. Squire (1996) Equity and growth in developing countries. Policy Research Working Paper No. 1563. Washington, DC: World Bank.

Burnside, C. and D. Dollar (1997) Aid, policies, and growth. Policy Research Working Paper No. 1,777. Washington, DC: World Bank.

Coate, S. and S. Morris (1996) Policy conditionality. Philadelphia: Department of Economics, University of Pennsylvania (In Mimeo).

Collier, P. (1996) 'The role of the state in economic development: Cross regional experiences', Paper presented at the AERC Plenary Sessions. Nairobi, Kenya, December.

Collier, P. and J. Gunning (1997) Explaining African economic performance. CSAE Discussion Paper. Oxford: Oxford University Press.

Collins, S. and B. Bosworth (1996) 'Economic growth in East Asia: Accumulation versus assimilation', *Brookings Papers on Economic Activity*, 2: 135–203.

Devarajan, S., A. A. Rajkumar and V. Swaroop (1999) What does aid do to African finance? Policy Research Paper. Washington, DC: World Bank.

Easterly, W. (1997) The ghost of financing gap: How the Harrod-Domar growth model still haunts development economics. Washington, DC: World Bank (In Mimeo).

Gordon, A. A. (1990) *Transforming Capitalism and Patriarchy*. Boulder and London: Lynne Rienner.

Griffin, K. B. (1970) 'Foreign capital, domestic savings and economic development', *Bulletin of the Oxford University Institute of Economics and Statistics*, 32 (2): 99–112.

Hansen, H. and F. Tarp (2000) The effectiveness of foreign aid. Development Economics Research Group, Institute of Economics, University of Copenhagen (In Mimeo).

Helleiner, G. (1993) 'External resource flows, debt relief and economic development in sub-Saharan Africa', in: A. Cornia and G. Helleiner, *From Adjustment to Development in Africa: Conflict, Controversy, Convergence and Consensus?* ch. 15. New York: St. Martins Press.

Heller, P. S. (1975) 'A model of public fiscal behavior in developing countries: Aid, investment, and taxation', *American Economic Review*, 65 (3): 429–445.

Hyden, G. (1983) *No Shortcuts to Progress: African Development Management in Perspective.* Heinemann, London, and University of California Berkeley.

——— (1986) 'The Anomally of the African Peasantry', *Development and Change*, 17 (4): 677–704.

Isham, J., D. Kaufmann and L. H. Pritchett (1997) 'Civil liberties, democracy, and the performance of government projects', *World Bank Economic Review*, 11 (2): 219–242.

Isham, J., D. Narayan and L. H. Pritchett (1995) 'Does participation improve performance? Establishing with subjective data', *World Bank Economic Review*, 9 (2): 175–200.

Kanbur, R. (1999) 'Prospectives and retrospective conditionality: Practicalities and fundamentals', in: P. Collier and C. Pattilo (eds), *Investment and Risk in Africa.* Basingstoke: Macmillan.

Killick, T. (1995) 'Conditionality and the adjustment-development connection', *Pakistan Journal of Applied Economics*, 11 (Summer/Winter): 17–36.

Lesink, R. and H. White (1999) *Aid Dependence: Issues and Indicators*. Stockholm: Expert Group on Development Issues (EGDI), Ministry of Foreign Affairs, Sweden.

Mosley, P. (1980) 'Aid, savings and growth revisited', *Oxford Bulletin of Economics and Statistics*, 42 (2): 79–95.

——— (1987) *Overseas Aid: Its Defence and Reform*. Brighton: Wheatsheaf Books.

Mosley, P., J. Hudson and S. Horrell (1987) 'Aid, the public sector and the market in less-developed countries', *Economic Journal*, 97 (387): 616–641.

Ndulu, B. J. (1986) 'Governance and economic management', in: R. Berg and J. Whitaker (eds), *Strategies for African Development*. Berkeley: University of California Press.

——— (1994) 'Foreign resource flows and financing of development in sub-Saharan Africa', in: *The International Monetary and Financial System: Developing Country Perspectives, International Monetary and Financial Issues for the 1990s* (vol. 4), G-24. Geneva: United Nations Conference on Trade and Development.

——— (1995) *Aid Dependence in Africa: Some Thoughts on the Nature of the Problem; Challenges for Transition to Less Dependence on AID*. Nairobi: African Economic Research Consortium.

Ndulu, B. and S. A. O'Connell (1999) 'Governance and growth in sub-Saharan Africa', *Journal of Economic Perspectives*, 13 (3): 41–66.

OECD (1998a) 'Geographical Distribution of Financial Flows to Aid Recipients', CD-ROM. Paris: Organization for Economic Co-operation and Development.

——— (1998b) 'Review of the International Aid System in Mali', Paper presented at the Special Meeting on the Mali Aid Review, 2–3 March, Paris. Organization for Economic Co-operation and Development.

Olson, M. (1982) *The Rise and Decline of Nations: Economic Growth, Stagflation, and Social Rigidities*. New Haven: Yale University Press.

Ravallion, M. and S. Chen (1997) 'What can new survey data tell us about recent change in distribution and poverty? *World Bank Economic Review*, 11 (2): 357–382.

Riddell, R. (1996) *Foreign Aid Reconsidered*. London: James Curry.

Sandbrook, R., and J. Barker (1985) *The Politics of Africa's Economic Stagnation*. Cambridge: Cambridge University Press.

Soludo, C. and S. A. O'Connell (1998) 'Aid intensity of Africa', Paper prepared for AERC/ODC collaborative research project on managing the transition from aid dependency in sub-Saharan Africa, October.

White, Howard and J. Luttik (1994) The countrywide effects of aid. Policy Research Working Paper 1337. Washington, DC: World Bank.

World Bank (1998) *Assessing Aid: What Works, What Doesn't, and Why*. New York: Oxford University Press.

Wuyts, M. (1996) 'Foreign aid, structural adjustment, and public management: The Mozambican experience', *Development and Change*, 27 (4): 717–49.

9 Aid, the Employment Relation and the Deserving Poor: Regaining Political Economy

Marc Wuyts

Introduction

As Kurt Martin argued, the pioneers of development economics saw the 'employment problem' as one of the central issues – if not the central issue – of development (1991: 37). His own seminal 1945 contribution, for example, took the absorption of labour surplus – rather than a specific rate of output – as the principal goal. More generally, in the spirit of classical political economy, the early tradition in development economics situated the absorption of the labour surplus within a context of economic transformation characterized by the process of industrialization. This specific focus was reflected in the use of concepts like the dual economy, disguised or hidden unemployment, the unlimited supply of labour and the creation and expansion of (formal) wage employment.

Yet, particularly in the last two decades of the twentieth century, there has been a marked shift away from these early concerns and from the concepts which embedded these concerns in theory. Therefore, while the pioneers of development economics talked about employment and growth, nowadays the new 'international consensus' talks about pro-poor growth. In recent years, then, poverty rather than unemployment (and the employment relation) has come to be the key concern.

This recent transition from unemployment to poverty is perhaps not all that surprising. Indeed, it could be seen as a normal progression over time as new or deeper concerns come to the fore which old concerns (and the concepts that went with them) failed to address. Yet this transition nevertheless becomes surprising and interesting once we take note that, about a hundred years earlier, the opposite transition took place: economic discourse shifted from poverty (or, paupers – as they were then called) to unemployment. In fact, the last decade of the nineteenth century and the first decade of the twentieth century witnessed the birth of *unemployment* as a new variable in social and economic analysis (Desrosières 1998: 254–259).

This paper takes a closer look at the substance and the reason for this historic conceptual reversal in economic discourse from poverty to

V. FitzGerald (ed.), Social Institutions and Economic Development, 169–187.

unemployment and then from unemployment back to poverty. It further links this latter transition with the accompanying changing emphases in foreign aid – that is, with the transition from aid as investment support to aid as poverty alleviation. Finally, the paper argues that this reversal from unemployment back to poverty needs to be situated within the broader process of what Makoto Itoh (2001: 110–124) aptly referred to as a 'spiral reversal' in capitalist development from the 1980s onwards. The approach adopted here, like that of Kurt Martin, is rooted in classical political economy and draws inspiration not only from the literature of the early pioneers of development economics and the debates surrounding aid and structural adjustment policies, but also from the recently emerging literature on the history of the theory and practice of social statistics (Stigler 1986, Kruger et al. 1987, Hacking 1990, Klein 1997, Desrosières 1998).

This paper first deals with the earlier transition from poverty to unemployment as the central concern of economic discourse for public policy in England in particular, and in the industrialized world in general, during the late nineteenth and early twentieth centuries. It then turns to development economics and looks at the reverse transition, from unemployment and the absorption of the labour surplus to poverty as the key concern of economic policy in developing countries in the late twentieth century. Finally, it shows how the changing emphases in foreign aid – the transition from aid as investment support to aid as poverty eradication – reflect this broader process of transition from the centrality of the employment relation back to poverty.

From poverty to unemployment: The emergence of a new variable

Unemployment as a concept is of relatively recent origin. In English, for example, the word only came in general use around the mid-1890s, long after industrial capitalism had taken hold, particularly in England (Garraty 1978 as quoted in Atkinson 1999: 67). It first appeared in the US Department of Labor Bulletin only in 1913 (*ibid.*). In public debates during the nineteenth century, it was pauperism not unemployment that occupied centre stage.

This obsession with paupers went back a long way. As De Swaan (1989: 17) points out, the debates on poor relief in early modern Europe invariably revolved around the issue of separating the 'deserving' from the 'undeserving' poor. In England in particular, the debate on pauperism came to focus on the specific question of whether poor relief was at all effective, or instead made matters worse by increasing poverty. Thomas Malthus spearheaded the latter position and fought relentlessly for the abolition of the then existing poor laws, based on parish relief.[1]

The new Poor Law of 1835 set up a dual system consisting of indoor and outdoor relief, both administered at the county level by the 'poor law unions'. In-relief took place in the workhouse where the able-bodied 'paupers' (mainly men) were housed in deplorable living conditions and forced to work at badly paid wages (the principle of 'least eligibility'). Out-relief consisted of the comparatively less harsh system of outdoor assistance administered mainly to women, the old, the disabled and the sick. The proportion of relief given out of doors was in fact seen as a reflection of the (relative) strictness or leniency of the management style of each local union (Desrosières 1998: 256, 133).

Subsequent political debates in England during the remainder of the nineteenth century evolved around the costs and effectiveness of both these systems in general, and of out-relief in particular. With respect to the analytical underpinnings of these debates, two authors, Charles Booth and George Udny Yule, are of particular relevance. Their respective contributions are important for two reasons. First, the issues they raised, both in content and in method, provided an early example of an analysis of the effectiveness of aid – albeit, in this case, of poor relief within national boundaries. Second, Booth's work in particular laid the analytical foundations for the transition from paupers to unemployment as the key concern of public debate.

As to the debate on the effectiveness of poor relief, Booth, in his seminal work, *The Aged Poor*, published in 1894, claimed that the proportion of relief given out of doors bore no relation to the total percentage of pauperism (quoted from Stigler 1986: 346). This assertion went counter to the widespread established belief – that is, among the rich and powerful – that out-relief was likely to worsen poverty. Yule, one of the most creative statisticians at the time, set out to investigate this claim made by Booth, using data for each of the 580 then-existing poor law unions. In the process, Yule not only broke new ground in the development of statistical theory, but also set the stage for an empirical approach to socio-economic analysis which economists have continued to rely upon up to today.

Yule took the total number of poor people receiving relief (outdoors plus indoors) to be a measure of the extent of pauperism, thus the dependent variable, and the ratio of out-relief to in-relief (i.e. the welfare to work-relief ratio) as an indicator of strictness of local union management, thus his independent variable (Desrosières 1998: 133–134, Stigler 1986: 347). Using the method of regression and correlation developed by Galton and Pearson, he first correlated both variables and found the coefficient of correlation to be positive and significantly different from zero (a value of 0.388 with probable error of 0.022) (Desrosières 1998: 134).

In subsequent work, Yule then continued to develop the concepts of partial and multiple regression so as to be able to look at the statistical correlation between his key variables, while controlling for other factors such as the proportion of elderly persons, average wage income, and differences in population (Desrosières 1998: 134, Stigler 1986: 355–356). The crux of his empirical analysis across his subsequent contributions led him to re-affirm his initial conclusion that "a high pauperism corresponds on the average to a high proportion of out-relief" (Yule as quoted in Klein 1997: 226). From his use of multiple regression, Yule further concluded that changes in pauperism were due to changes in administrative policy and not to external causes such as the growth of population or economic changes (ibid.).

Yule's development of statistical method was a veritable *tour de force*. In technical terms, the trajectory of the development of his work proceeded as follows: Yule first applied simple regression analysis as developed by Galton and Pearson on the basis of the bivariate normal (probability) distribution. Application of this technique, therefore, assumed that the data satisfied the normality assumption. Yule was aware, however, that his data (like most social and economic data) were skewed and, hence, that the normality assumption was unwarranted. In subsequent theoretical work, Yule went back to Legendre's older 'principle of least squares'(formulated in 1805), connected it with Galton and Pearson's concepts of regression and correlation (developed in the mid-1880s), dropped the normality assumption and thus redefined regression as curve-fitting using least squares. Finally, Yule then generalized his newly found method by developing multiple and partial regression using the method of least squares[2] (Stigler 1986: 346–358).

Like many statisticians and econometricians after him, however, Yule tended to be more enchanted by the intricacies of statistical method than by the problem at hand.[3] In this particular case, the key issue was how poverty itself was measured. Indeed, the data used in assessing the extent of local poverty (Yule's dependent variable and performance indicator) and the means used to relieve it, all derived from the same source – the poor law unions administering the relief (Desrosières 1998: 256). In other words, the extent of poverty was defined by the measures taken to fight it. Hence, *ceteris paribus*, the more lenient the poor administration authority, the more inclusive it was in administering relief, and hence, the greater the number of 'paupers'. The definition of who was a pauper, therefore, depended on the leniency of poor relief. This problem of 'circularity', Desrosières points out, "seems to have concerned neither Yule nor those to whom he presented his study" (ibid.).

Charles Booth's work took a different track and managed to circumvent the problem of circularity. To do so, he relied on own survey data to analyse the nature of urban poverty in London.[4] The power of Booth's approach was that it drew attention to the characteristics of family income – both its level and its regularity – and used this to identify different types of poverty. This approach broke decisively with the then-existing practice of dividing the lower classes into three ensembles: the dangerous classes, the poor, and workers in general. Instead, Booth used eight hierarchical categories to capture the varied patterns in the sustenance of urban livelihoods. The very poor included the 'infamous' whose sources of income were dishonest or unknown on the one hand, and those families relying on casual work in a state of chronic destitution on the other. The poor were subdivided into two categories depending on the regularity of their incomes: those with intermittent incomes as a result of the vagaries of seasonal unemployment and those with low but stable incomes. These groups of the poor and the very poor were then distinguished from the 'comfortable working class' (including the borderline category of workers with regular incomes minimally sufficient to afford a living) and the 'lower and upper middle classes' (Desrosières 1998: 257).

Francis Galton – aristocrat, founder of the eugenics movement and inventor of the statistical concept of regression – was quick to grind Booth's categories to his own mill by ordering them along a continuous variable of 'civic worth', distributed in accordance with the normal distribution (ibid.: 114). In other words, for Galton, poverty was a state of affairs characteristic of certain types of people – the inferior classes. The real distinctive feature of Booth's work, however, was its premise that poverty and inequality derive from the condition of (wage) employment. This is evident from the importance he accorded to the character of labour regimes, and the level and regularity of incomes they supported, in determining the level of economic security of families. This aspect of Booth's work prepared the ground for a "transition from the old idea of poverty to the as yet non-existing idea of unemployment – the temporary loss of a wage-earning position that guaranteed a regular income" (ibid.: 257).

The political conditions that made this transition possible came about with the rise of a new generation of social reformers who expressed "the problems of poverty in terms of regulating the labour market and passing laws to provide social protection, rather than relying on local charity" (Desrosières 1998: 262–263). Indeed, as Amartya Sen (1981: 173) points out, "the phase of economic development *after* the emergence of a large class of wage labourers but *before* the development of social security arrangements is potentially a deeply vulnerable one". In the context of the so-called 'sweating system', where wage labour was unregulated, flexible and insecure, this vulnerability was essentially hidden within the swamp of pauperism. In this system, work was subcontracted by employers to intermediaries (subcontractors) who then recruited the necessary labour force. No formal bond existed between employer and worker (Desrosières 1998: 263). Labour was mobilised whenever work was available and only for as long as it lasted without any formal ties between worker and employers.

For unemployment to emerge on the scene as a new variable, therefore, there had to be an employment relation that could be formally broken such that unemployment results. In other words, as Atkinson (1999: 68) puts it, "unemployment is associated with a labour market situation where employment is a 0/1 phenomena". It was the large-scale factory, assembling masses of (mainly male) workers separated in time and space from their families, that provided the main impetus for the development of an employment relation within which the modern concept of unemployment could arise (ibid.). It was also this process that fostered trade-union formation, led to the first records being kept on unemployment (within the unions and friendly societies) and propelled social struggles for the improvement of the working classes. When laws were passed to regulate the position of the wage-earning workforce and the duties of the employers, unemployment could then be defined and came to be measured as a break in the bond between workers and their bosses (Desrosières 1998: 263). Consequently, it became possible to demarcate a dividing line between the duties of the employer in providing regular employment and the need for social security when the employment relation was broken. Social struggles thereafter, thus took on a different dimension: welfare not poverty was to be associated with large numbers of people (Metz 1987: 347).

This shift in emphasis from poverty to unemployment took place against a background in which, for about a century from the late nineteenth century onwards, the capitalist world system moved away from competitive free market capitalism (Itoh 2001: 113). This spiral reversal away from the earlier nineteenth century liberalism in economic development had some notable characteristics:

- concentration of mass numbers of workers in large factories and the impulse this gave to the development of trade unions,
- introduction of Taylorism as a new organization of labour that allowed rapid and continuous gains in productivity,
- progressive development of a regime of accumulation coupled with a mode of regulation where the gains of productivity growth were systematically redistributed to every social class,
- development of a network of collective bargaining, social legislation and, ultimately, the welfare state (Itoh 2001: 113–115, Lipietz 2001: 18).

This era, in its mature form, came to be characterized alternatively as Keynesianism or as Fordism. It was the prior emergence of the concept of unemployment that paved the way for the rise of Keynesian economics during the great depression of the 1930s, when unemployment rose to unprecedented levels. Concern with the dynamics of unemployment thus came to occupy centre stage. Keynes emphasized the importance of state action 'from above' through deliberate fiscal and monetary policies to boost effective demand in order to sustain full employment. In contrast, Fordism emphasized the rise in real wages in line with productivity growth by means of explicit or implicit capital-labour agreements to enable the growth in effective demand to keep pace with the growth in output of large-scale industry (Itoh 2001: 115–117, Lipietz 2001: 18–21).

In sum, then, the transition from poverty to unemployment meant that poverty ceased to be seen as a condition of certain people (their civic worth as Galton put it), but instead as reflective of the nature of the employment relation within capitalist development. Social struggles thereafter came to revolve around the conditions of wage employment and the social protection of the unemployed. Pauperism as a concept receded to the background. Inequality came to be addressed as a condition of (wage) labour and not as deficiency of people.[5]

The reverse transition: From unemployment back to poverty

Development economics emerged as a separate discipline in the wake of World War II, during the heyday of Keynesianism. The pioneers of development economics realized, however, that the problem of effective demand was not the principal issue for development (Rao [1952] 1958). Developing economies, they argued, were supply-constrained, not demand-constrained. But this did not mean that they were not deeply influenced by the economic discourse at the time and by its focus on the macroeconomics of employment and the dynamics of unemployment. On the contrary, they acknowledged that in developing countries large-scale hidden or disguised unemployment prevailed, but argued that this problem could not be remedied by boosting effective demand. Instead, what was needed, they argued, was a protracted transformation of the developing economy to absorb surplus labour through the expansion of wage labour in the process of industrialization fuelled by investment.

Wage employment, as Kurt Martin argued (1991: 36–37), was seen by the founders of development economics as a "historical concept and phenomenon" and, hence, as far as developing countries were concerned, as a process in the making. The main analytical tool they employed to look at this process was twofold: the notion of the 'dual economy' characterized by the juxtaposition of an emerging 'modern' sector with a pre-existing 'traditional' sector, and the concept of disguised or hidden unemployment. The latter, as Martin explained, was

> defined as a situation in the productivity dimension, and refers to people not normally engaged in wage employment who can be transferred into more

> modern, rewarding activities without loss of output in the so-called traditional sectors – usually, but not exclusively, peasant agriculture (1991: 30).

Their approach to development economics was macroeconomic in scope and the focus was on the interrelationships between the regime of accumulation, productivity growth and the real wage in the process of industrialization. The Lewis model best typified these concerns. Capital accumulation in industry (or the modern sector at large) was assumed to take place by drawing surplus labour from agriculture (or, more generally, from the traditional sector) at a real wage rate somewhat above the subsistence level prevalent in the traditional sector. Productivity growth in the modern sector would fuel profits, thereby allowing for the acceleration of capital accumulation. Once surplus labour dried up, the real wage was assumed to rise presumably in line with productivity.

The argument, as briefly stated above, says little about the corresponding mode of regulation inherent in the prevailing regime of accumulation. In fact, it was the conviction of the post-war development economists that economic growth in the developing countries 'as a matter of economic logic' was bound to take the same general direction (towards industry and towards wage employment) as it had done in the developed countries (Martin 1991: 30). In so doing, however, their frame of reference was modelled on the prevalent capitalist developed economies and not on their more virulent liberal variants of the nineteenth century. The absorption of surplus labour from the traditional sector, therefore, was assumed to go in the direction of *formal* wage labour. In other words, using Booth's language, during the phase of the absorption of surplus labour, wage labour would hover at the borderline of the 'comfortable' working classes (low but secure wages and, presumably, further protected by minimal social security). Thereafter, average wages would increase within a context of secure employment. Consequently, the question of economic insecurity associated with the creation of wage labour initially did not feature much in their analysis.[6]

The strength of this approach lay in its focus on the analysis of the regime of accumulation and of the way this shaped the relation between productivity growth and real wages and between industry and agriculture. In so doing, the early pioneers stressed the importance of the specificity of the developing countries and the necessity to look at process in a context of transformational growth. Central to it all was the dynamics of employment – the transfer of labour from agriculture to industry and its transformation into wage labour. On this count, the early-tradition development economists were clearly marked both by their classical antecedents as well as by Keynesian macroeconomics and its emphasis on employment. For them, therefore, the dynamics of employment as determined by the regime of accumulation was the key to understand distribution and, hence, the prevalence of poverty in society.

From the 1980s onwards, however, this tradition in development economics came under severe attack under the impulse of the neoclassical resurgence. The concept of transition now came to mean a transition to open market economy – a reassertion of the earlier tradition of liberalism. The emphasis on specificity in analysis gave way to the premise of its universal validity. The focus on the macro foundations of economic development shifted towards concentration on the micro foundations of macro outcomes. The emphasis on market failure and its consequent call for state action was replaced by an emphasis on government failure. And, after the initial euphoria of market-led development had ceased to eclipse all other concerns, the emphasis shifted to poverty and away from the earlier focus on the dynamics of employment and unemployment. At the level of economic policy discourse, therefore, a complete turn-around took place in the matter of only a decade or so.

This change of gear in economic discourse was not confined solely to development economics, but reflected the broader challenge against Keynesian economics during the 1970s. These developments within economic discourses, however, did not take place in isolation. The broader context clearly related to the tapering off of the long post-World War II boom. The demise of economics' Keynesian and Fordist foundations (as well as the decline and subsequent demise of the Soviet experience) was reflected in what Itoh (2001) aptly describes as a *spiral reversal* in capitalist development back to a more virulent liberalism in a novel context of accelerated globalization.

This spiral reversal, which ruptured the earlier more uniform regimes of accumulation, was not replaced by a unique new model. Rather, it provoked a variety of responses – alternative paths – through which the industrialized and newly emerging industrial countries tried to adjust to the new and more competitive environment in search of alternative regimes of accumulation and modes of regulation (Itoh 2001, Lipietz 2001). In other words, as Itoh (2001: 122) put it, "the process of globalisation in our age is not necessarily homogenising the economic systems of the world". In the poorer developing countries, however, structural adjustment, with the financial backing of foreign aid, provided a new framework and model within which (admissible) policy discourses were set and actual adjustments were forged, a point returned to in the final section of this paper.

As argued earlier, the transition from poverty to unemployment took place against a background of societal transformations which regulated the condition of labour and its relation to capital within the process of accumulation. In similar vein, the initial processes of import-substituting industrialization in Third World countries were largely structured along a model of the creation and expansion of *formal* wage labour. This explains, for example, why the development of the so-called 'informal sector' was initially viewed as an 'aberration', rather than as the 'desired norm'. Structural adjustment, however, and its emphasis on a reversal to competitive free market capitalism fostered the deconstruction of pre-existing modes of regulation, including those that structured the formalization of wage labour. This process of deconstruction revealed itself in the gradual erosion of formal wage labour within Third World countries and in the expansion and spread of patterns of accumulation based on more flexible and insecure labour regimes – casual, seasonal or more secure wage labour, poorly or better paid, combined with self-employment using household labour, paid or unpaid (Wuyts 2001). Not surprisingly, then, in the era of structural adjustment, the 'informal sector' came to be heralded as an example of vibrancy in development rather than as an aberration.

Consequently, the earlier notion of the classical tradition in development economics that the *direction* of development would be towards the expansion of formal wage labour became increasingly questionable. Wage labour no longer evolved in the direction of a *crispy* set where employment is a 0/1 phenomenon (Atkinson 1999: 68). Instead it turned increasingly into a much more *fuzzy* set. Under the impulse of structural adjustment, therefore, employment became more amorphous in nature – varied, more insecure and differentiated, but hard to pin down. The process of the de-formalization of wage labour subsequently led to greater emphasis being put on poverty analysis within the tradition of the approach developed by Seebohm Rowntree in 1910, based on his poverty studies in York during the late nineteenth and early twentieth centuries (Kanbur and Squire 2001: 186).

This approach involved the measurement of poverty by means of a poverty line defined by the *socially acceptable* amount of money (income) required "to obtain the minimum necessaries for the maintenance of merely physical efficiency" (Rowntree [1910] as quoted in Kanbur and Squire 2001: 186). Using survey analysis based on household data, the poor are then identified as those falling below the poverty line and the extent of poverty is given by the estimated proportion of the population falling into this category. Subsequently, the characteristics of the poor can be analysed with respect to different factors – e.g. gender, location (rural/urban), educational background, occupation. This approach, developed nearly a century ago, is very much the main method by which poverty is measured today (ibid.: 216).

A recurrent conclusion – or stylized fact – of modern applications of such poverty studies is the observed marginalization, if not isolation, of the poor within market circuits and, hence, their non-existing or low access to sources of cash incomes, particularly in rural areas (Kanbur and Squire 2001: 206). Obviously to someone like Kurt Martin, steeped in the tradition of classical political economy, this finding would have come as no surprise since it does little more than provide empirical substance to Marx's concept of the 'latent surplus population' as part of the 'reserve army of labour' under capitalist accumulation. In fact, as Martin (1991: 30) argued, the concept of 'disguised' or 'hidden' unemployment, central to early development economics, was not an 'innovation', but instead a 'rediscovery' ("usually without realizing that it was a rediscovery"; ibid.) of Marx's earlier concept.

The novelty of the present-day mainstream analysis of poverty, however, is that it draws the opposite conclusion from this stylized fact. As argued, for example, by Collier et al. (1986: 77, 81, 94, 133), the problem of the poor is not that of 'too much market', but instead 'too little of it'. Consequently, poverty reduction is seen to result from widening access to markets and from the growth of the market economy, coupled with policy measures aimed to strengthen the enabling conditions of the poor (their human capital, access to credit, etc.). Rapid economic growth, achieved by opening up the economy, therefore, is identified as the main factor in poverty reduction (Collier 1999: 538).

There is, however, still much to be said for the old concern in political economy with the relation between capitalist development and the condition of labour, including its manifestation in open or disguised unemployment. To put these old concerns in a more modern dress, it is useful to draw upon A. Sen's distinction between the space of commodities and the space of capabilities. Sen (1984a: 335) argues that poverty is an absolute notion in the latter, but often takes a relative form in the former. Consequently, *inequality* in the commodity space (and, hence, in income as well) can engender absolute deprivation in the capabilities space. This is an important insight. But it leads to the further point that inequality, like capabilities, is inevitably contextual in nature and, hence, depends on the prevailing character of capitalist development.

This point reveals itself clearly, for example, in Sen's own analysis of famines. At the abstract level, Sen (1984b) talks about an individual, i, equipped with an endowment vector, x_i, and confronting a specific exchange entitlement mapping, $E_i(\bullet)$. Entitlement failures leading to deprivation and hunger are then defined in general terms, independent of the specific insertion of the individual within the organization of production and exchange. But when Sen goes on to analyse concrete cases of famines, his operational category is no longer the individual i, but the social group j consisting of individuals who share common characteristics, for example, landless labourers in Bangladesh. Social groups, particularly when it concerns the poor, are therefore defined mainly with respect to their shared conditions of labour within a specific regime of accumulation.

Wage labour, indeed, has become much more of a fuzzy set and, hence, poverty may increasingly appear as a 'seamless web' where the poor, faced with a great deal of risk, "try to diversify their vulnerability to disaster by diversifying their income" (Kanbur and Squire 2001: 199–200, 205). That does not mean, however, that the analysis of the dynamics of wage labour and its implications for the creation and maintenance of unemployment, open or disguised, ceases to be important. The relation between the regime of capitalist accumulation and its mode of regulation, on the one hand, and the process by which productivity growth is distributed among different classes (and contributes to the growth of real wages, in particular), on the other, still remains essential to understanding whether or not, and how, economic growth (if it occurs at all) translates into poverty reduction (Akyüz and Gore 2001, Karshenas 2001, Wuyts 2001).

This section has argued that the discourse in development economics has shifted away from unemployment (open or disguised) back to poverty as the key central concept. This shift means that inequality is no longer so much seen as the condition of labour inherent within a specific regime of accumulation, thereby shaping the nature of deprivation in capabilities. Instead poverty has become a property of individuals, and the creation of an enabling environment (including greater access to markets) is required for poverty to be overcome. The classic concern with the systemic dimension underlying the dynamics of poverty within capitalist development, therefore, appears to have been taken out of the equation. Yet as is argued below, at the level of policy discourse, the *systemic prescription* is nevertheless still very much on the agenda.

From aid as investment support to aid as poverty alleviation

Policy discourse in the poorer developing countries is heavily shaped by the politics of foreign aid. Figure 1 shows the broad changes in emphases in 'donor' approaches to foreign aid in the post-World War II period. As the figure illustrates, the emphasis has shifted from aid-financed investment to aid-induced policy reforms (Ali, Malwandi and Suliman 1999: 508), and more recently, to aid induced by policy reforms (Burnside and Dollar 1997, Dollar and Easterly 1999, Collier 1999). More broadly, there has been a noticeable shift away from aid as investment support to aid as poverty alleviation.

Aid as investment support in the form of project aid was the typical form of assistance during the heyday of traditional development economics. This form of aid inserted itself neatly within the then-prevailing overall policy framework of import-substituting industrialization based on the transfer of labour from agriculture to industry and its transformation into *formal* wage labour. Donor confinement to project aid meant that donors operated

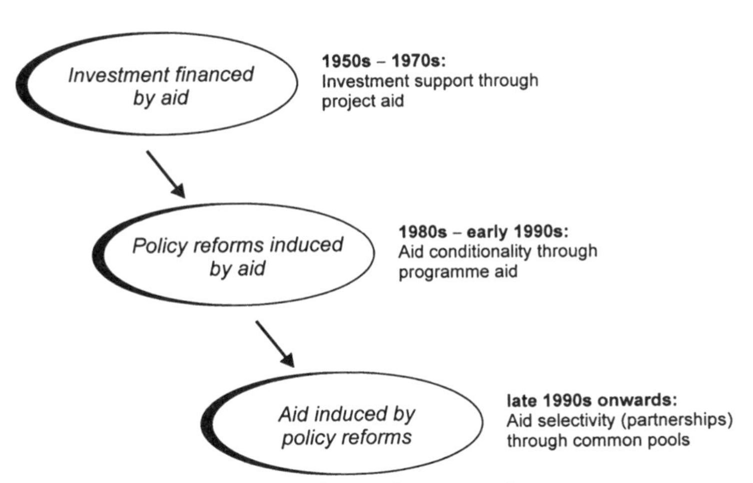

Figure 1 Changing (donor) approaches to foreign aid

at the micro level within an overall context in which economic policy was essentially steered by macro-structuralist concerns. Policy debates on aid mainly centred on whether aid fuelled investment, and whether investment in turn fuelled economic growth. The question of ownership and governance hardly arose within the literature at the time. The reason why it did not, as Mkandawire (1994: 165) points out, was that in these early post-colonial days, donors in general, and the World Bank in particular, viewed the state "within a basically modernisation paradigm, in which the state played an essentially benign, developmental role". In any case, Mkandawire adds, "the bank's confinement to projects did not necessitate a more complex view of the state than this" (ibid.).[7]

Under structural adjustment in the 1980s aid policies shifted dramatically as programme aid entered the scene. Structural adjustment was essentially seen as a quick-fix operation to effect a rapid transition to an open economy (the big-bang notion), backed by exceptional finance subject to policy conditionality (Toye 1994). Initially, there was little explicit concern for social development and poverty reduction since the sheer force of opening up developing economies to market forces was seen as the main mechanism to foster economic growth and, hence, to reduce poverty. Programme aid as quick-disbursing finance was meant to cushion the blow of initial reforms and permit a rapid transition to market-based economic growth, propelled by improvements in the policy environment. Aid conditionality was the vehicle through which rapid changes in the policy environment could be forged. Hence, it was meant to yield 'value for money' for the large amounts of exceptional finance made available.

At the political level, these aid-induced economic reforms were accompanied by two major changes. First, the state was no longer seen as the main agent of change, but instead, as part of the problem because of its assumed predatory, interest-seeking and rent-protecting nature. As Toye points out, the 'urban bias' thesis played an important role in shifting donors' perceptions of the state in developing countries (1994: 25). Indeed, the thesis entailed that both efficiency and equity were jeopardized because urban interests dominated the political scene. Consequently, as Mkandawire (1994: 165) points out, the urban bias thesis explained "why it was politically rational to persist with what were obviously irrational policies". This argument, pushed to its logical conclusion, had a strong "authoritarian predilection" and "anti-political penchant" (ibid.: 166–169). Consequently, democratic sensibilities were on a rather low ebb as far as the donor community in general, and the World Bank in particular, were concerned (Toye 1994: 26). Instead, policy changes were to be propelled by aid conditionality. The aid relation, therefore, came to be seen as a *principal-agent* problem where the leader, the donor, relies on aid conditionality to enforce compliance by the follower, the recipient state.

Second, the move by donors onto the macro scene led to the formation of the 'donor community' as an entity distinct from each individual donor and with a dominant, if not overriding voice in the domestic policy discourses of the recipient countries. Hence, at a time when the domestic state was pushed back into the defensive, donors gained voice as a community by creating a common platform to exert leverage on local states.

Both factors, taken together, meant that during the 1980s there was as yet little concern on the part of (most) donors with the link between democratization and economic reforms, nor with the questions of 'ownership' and 'governance' that were to appear later onto the scene. It was only in the early 1990s, when it became increasingly obvious that structural adjustment did not provide the quick-fix solution initially envisioned, that donor concerns shifted towards greater concern with social development and, later, poverty alleviation on the one hand and with ownership and governance on the other.

The econometrics of Burnside and Dollar (1997) played an important role in shifting aid policies away from aid conditionality towards *aid selectivity* in the context of partnerships.[8] There is, however, an interesting parallel between the work of Burnside and Dollar in the mid-1990s and that of Yule a century earlier. As discussed earlier, Yule sought to show that, controlling for third factors, relief worked when administered indoors, but not when given outdoors. Substituting 'indoors' for 'good policies' and 'outdoors' for 'inappropriate policies' leads to the conclusion arrived at by Burnside and Dollar, but this time within a cross-country setting. In other words, aid works when good policies are in place and, hence, aid should be concentrated on those countries where the policy environment is receptive.[9]

There is, therefore, a 'new consensus' emerging as to the nature of 'good policies', a particular mix of market openness, concern for poverty alleviation, democracy and good governance (see e.g. Herfkens 1999). The stress on partnership and ownership appears as a welcome change from the earlier practice of imposed conditionality, which essentially boiled down to 'informal governance' by international institutions (Mkandawire 1994: 162). But, as Mkandawire (ibid.: 173) further points out, there also appears to be an implicit assumption that the economic prescriptions underlying the consensus are in fact the only way out and, hence, that democratic processes within developing countries will inevitably converge on such policies.

As Itoh (2001) notes, however, present-day developments within the industrialized world are far from homogenous in terms of their respective directions. With respect to social development, for example, Sen (1999: 95), contrasting Western Europe and the United States, points out that, while Western Europeans would find it hard to accept the lack of social provisioning in the United States, US citizens would find the double-digit levels of unemployment in Europe quite intolerable. Significant divergences in policies exist, therefore, between Europe and the United States, particularly in terms of how their respective regimes of accumulation structure the relation between productivity growth and real wages (including social provisioning, protection and exclusion). In similar vein, Taylor, Mehrotra and Delamonica (1997) argue that historically the articulation and synergy between economic growth, social development and poverty reduction in the Third World show a great variety of patterns, both in their successes and failures.

This suggests the need for recognizing the importance of specificity in process and context as dependent on past experiences in economic development. Yet in contrast, the 'good policies' advocated by the new international consensus on development aid are remarkably uniform in their essence. Genuine partnership, however, must come to terms with differences in perspectives on viable ways to achieve economic growth and poverty reduction and give room to partners to learn from mistakes. The danger of aid selectivity based on a donor-driven consensus resides in the 'excessive levelling' of the economic and political landscapes that it engenders in the poorer developing countries (Mkandawire 2001: 310). The shift from aid as investment support to aid as poverty alleviation entails the danger, therefore, of a return to the earlier nineteenth-century inspired politics of the 'deserving poor', where richer nations effectively dictate the limits within which the 'deserving' poorer ones can operate.

Notes

1. As a consequence, 'demography became the numerical study of poverty'. The execution of the first census in 1801, for example, was made the responsibility of the offices of the poor laws. The offices of the poor laws were indeed the only existing official institutions capable of handling large-number phenomena (Metz 1986: 343).
2. Not surprisingly, the momentum of least squares regression, thus unleashed by Yule, proved to be unstoppable. As Klein explains: Economists, in particular, latched onto least sum of squares estimation that enabled them to mimic the *ceteris paribus* properties of a laboratory experiment, draw law curves in logical time, and estimate inexact relationships between variables with skewed distributions. (Klein 1997: 226).
3. As Desrosières (1998: 134) put it, reading Yule's work, "one could easily get the impression that his first concern was to demonstrate the interest offered by the new tools, while at the same time putting his own stamp on them".
4. Seeborn Rowntree's 1899 poverty study in York followed a similar tract and would eventually lead him to pioneer the method of poverty lines as the amount of money required to obtain the minimum necessities of life (Sen 1984a: 326, Kanbur and Squire 2001: 186).
5. Poverty lines as pioneered by Rowntree played an important role in setting the standard for welfare policies. In the context of these policies, the aim was to improve the condition of labour within capitalist development shaped by Keynesianism and Fordism.
6. The rapid rise in urban unemployment in Third World countries and the growth of what came to be known as the 'informal' sector (where prevailing labour relations were akin to those described by Charles Booth) nevertheless brought the issue of economic security more and more to the fore. These developments fitted uneasily within the confines of the established doctrine and were catered for by looking at unemployment and urban informal sector employment as shock absorbers between rural out-migration and formal sector employment.
7. This did not mean, however, that neo-liberalism was completely dormant. On the contrary, as Toye (1994: 24) points out, "the great neo-liberal project of the 1970s had been the use of shadow prices in project appraisal in order to secure the more rational use of public sector investment resources". More specifically, "shadow prices was a method for trying to adjust the public part of the economy, while allowing the macro framework itself to remain out of line with international economic forces". In other words, the main project of the World Bank at the time was 'getting shadow prices right'.
8. Aid conditionality implies that aid is given conditional upon the adoption of a set of specific policy measures. Aid selectivity, in contrast, concerns donor selec-

tivity in the choice of the country (partner) to which aid is given. The basic idea is that aid is to be concentrated on like-minded partners whose policy environment is seen to be receptive of aid.

9. A discussion of the econometrics involved is beyond the scope of this paper. A useful reference in this respect is Hansen and Tarp (2000). Interestingly, the econometrics of Burnside and Dollar are not devoid of some element of 'circularity'. Indeed, the choice of the key *independent* variable, a composite indicator of 'good policies', appears to be somewhat guided by what works in terms of getting the right results.

References

Abritton, R., M. Itoh, R. Westra and A. Zuege (eds) (2001) *Phases of Capitalist Development: Booms, Crises and Globalizations*. Houndsmills: Palgrave.

Akyüz, Y. and C. Gore (2001) 'African economic development in a comparative perspective', *Cambridge Journal of Economics*, 25 (3) May: 265–288.

Ali, A. A. G., C. Malwanda and Y. Suliman (1999) 'Official development assistance to Africa: An overview', *Journal of African Economies*, 8 (4): 504–527.

Atkinson, A. B. (1999) *The Economic Consequences of Rolling Back the Welfare State*. Munich Lectures in Economics. Cambridge, Massachusetts: MIT Press.

Burnside, C. and D. Dollar (1997) Aid, policies and growth. Policy Research Working Paper 1,777. Washington, DC: World Bank.

Chenery, H., M. S. Ahluwalia, C. L. G. Bell, J. H. Duloy and R. Jolly (1974) *Redistribution with Growth*. Oxford: Oxford University Press.

Collier, P., S. Radwan and S. Wangwe with A. Wagner (1986) *Labour and Poverty in Rural Tanzania. Ujamaa and Rural Development in the United Republic of Tanzania*. Oxford: Clarendon Press.

Collier, P. (1999) 'Aid dependency: A critique', *Journal of African Economies*, 8 (4): 528–545.

Desrosières, A. (1998) *The Politics of Large Numbers: A History of Statistical Reasoning*. New Haven: Harvard University Press.

De Swaan, A. (1988) *In Care of the State*. Oxford: Polity Press.

Dollar, D. and W. Easterly (1999) 'The search for the key: Aid, investment and policies in Africa', *Journal of African Economies*, 8 (4): 546–577.

Hácking, I. (1990) *The Taming of Chance*. Cambridge: Cambridge University Press.

Hart, G. (1995) 'Clothes for next to nothing: Rethinking global competition', *SA Labour Bulletin*, 19 (6) December: 41–47.

——— (1996) 'The agrarian question and industrial dispersal in South Africa: Agro-industrial linkages through Asian lenses', *Journal of Peasant Studies*, 23 (2/3) January/April: 245–277.

Herfkens, E. (1999) 'Aid works – let's prove it!', *Journal of African Economies*, 8 (4): 481–486.

Hansen, H. and F. Tarp (2000) 'Aid effectiveness disputed', in: F. Tarp with P. Hjertholm (eds), *Foreign Aid and Development. Lessons Learnt and Directions for the Future*, pp. 103–128. London & New-York: Routledge.

Itoh, M. (2001) 'Spiral reversals of capitalist development: What does it imply for the twenty-first century?' in: A. Abritton et al. (eds), *Phases of Capitalist Development: Booms, Crises and Globalizations*, pp. 110–124. Houndsmill: Palgrave.

Itoh, M. and C. Lapavitsas (1999) *Political Economy of Money and Finance*. Houndsmill: Macmillan.

Kanbur, R. and L. Squire (2001) 'The evolution of thinking about poverty: Exploring the interactions', in: G. M. Meier and J. E. Stiglitz, *Frontiers of Development Economics: The Future in Perspective*, pp. 183–226. Oxford: Oxford University Press and World Bank.

Karshenas, M. (2001) 'Agriculture and economic development in sub-Saharan Africa', *Cambridge Journal of Economics* 25 (3) May: 343–368.

Klein, J. L. (1997) *Statistical Visions in Time: A History of Time Series Analysis 1662–1938*. Cambridge: Cambridge University Press.

Kruger, Lorenz, Lorraine J. Daston and Michael Heidelberger (1987) *The Probabilistic Revolution, Vol. 1: Ideas in History*. Cambridge, MA: MIT Press.

Lipietz, A. (2001) 'The fortunes and misfortunes of post-Fordism', in: A. Abritton et al. (eds) *Phases of Capitalist Development: Booms, Crises and Globalizations*, pp. 17–36. Houndsmill: Palgrave.

Martin, K. (1991) 'Modern development theory', in: K. Martin (ed.), *Strategies of Economic Development: Readings in the Political Economy of Industrialization*, pp. 27–74. Houndsmill: Macmillan (with the Institute of Social Studies).

Metz, K. H. (1987) 'Paupers and numbers: The statistical argument for social reform in Britain during the period of industrialization', in: L. Krüger, L. J. Daston and M. Heidelberger (eds), *The Probabilistic Revolution. Volume 1: Ideas in History*, pp. 337–350. Cambridge, MA: MIT Press.

Mkandawire, T. (1994) 'Adjustment, political conditionality and democratisation in Africa', in: G. A. Cornia and G. K. Helleiner (eds), *From Adjustment to Development in Africa: A UNICEF Study*, pp. 155–173. Houndsmill: Macmillan.

——— (2001) 'Thinking about the developmental states in Africa', *Cambridge Journal of Economics*, 25 (3) May: 265–288.

Rao, V. K. R. V. ([1952] 1958) 'Investment, income and the multiplier in an underdeveloped economy', in: A. N. Agarwala and S. P. Singh, *The Economics of Underdevelopment*, pp. 205–218. Delhi: Oxford University Press.

Sen, A. K. (1981) *Poverty and Famines. An Essay in Entitlement and Deprivation*. Oxford: Clarendon Press.

——— ([1983] 1984a) 'Poor, relatively speaking', in: A. Sen, *Resources, Values and Development*, pp. 325–345. Oxford: Blackwell.

——— ([1983] 1984b) 'Ingredients of famine analysis: Availability and entitlement', in: A. Sen, *Resources, Values and Development*, pp. 452–484. Oxford: Blackwell.

——— (1999) *Development as Freedom*. New York: Alfred A. Knopf.

Stigler, S. M. (1986) *The History of Statistics: The Measurement of Uncertainty before 1900*. Cambridge, MA: Belknap Press of Harvard University.

Taylor, L., S. Mehrotra, and E. Delmonica (1997) 'The links between economic growth, poverty reduction, and social development: Theory and policy', in: S. Mehrotra and R. Jolly, *Development with a Human Face: Experiences in Social Achievement and Economic Growth*, pp. 435–468. Oxford: Oxford University Press.

Toye, J. (1994) 'Structural adjustment: Context, assumptions, origin and diversity', in: Van der Hoeven and F. Van der Kraaij, *Structural Adjustment and Beyond in Sub-Saharan Africa: Research and Policy Issues*, pp. 18–35. The Hague: Ministry of Foreign Affairs, in association with J. Currey (London) and Heinemann (Portsmouth).

Wuyts, M. (2001) 'Informal economy, wage goods and accumulation under structural adjustment: Theoretical reflections based on the Tanzanian experience', *Cambridge Journal of Economics*, 25 (3) May: 417–438.

Contributors

Valpy FitzGerald is Reader in International Economics and Finance and Director of the Finance and Trade Policy Research Centre at the University of Oxford, UK. He is also Professor of Economics affiliated with the Institute of Social Studies, The Netherlands.

Richard R. Nelson is Professor in the School of International and Public Affairs, Business and Law at Columbia University, USA.

J. B. Opschoor is Professor of Development Studies and the current Rector of the Institute of Social Studies. He is also Professor of Environmental Economics at the Free University of Amsterdam, The Netherlands.

José Antonio Ocampo is Executive Secretary of the United Nations Economic Commission for Latin America and the Caribbean, CEPAL/ECLAC, Chile.

Rob Vos is Professor of Finance and Development at the Institute of Social Studies, The Netherlands.

Frances Stewart is Professor of Development Economics and Director of the International Development Centre, University of Oxford, UK.

B. J. Ndulu is Sector Lead Specialist at the Macroeconomics, World Bank Country Office, Tanzania.

Marc Wuyts is Professor of Applied Quantitive Economics at the International Institute of Social Studies, The Netherlands.

Index